FULL AUTO

Modification Manual

AF826177

Desert Publications
El Dorado, AR 71731 1751 U. S. A.

Full Auto

Modification Manual

Published by Desert Publications
505 East Fifth Street
El Dorado, AR 71730 U.S.A.
1-870-862-3811
info@deltapress.com

Printed in U.S.A
ISBN: 978-0-87947-166-2
10 9 8 7 6 5
Desert Publication is a division of
The DELTA GROUP, Ltd.

Direct all inquiries & orders to the above address.

Neither the author nor the publisher assumes any responsibility for the use or misuse of the information contained in this book. This material was compiled for educational and entertainment purposes and one should not construe that any other purpose is suggested.

Warning!

The Publisher (Desert Publications) produces this book for informational and entertainment purposes and under no circumstances advises, encourages or approves of use of this material in any manner.

WARNING NOTICE

Neither the author nor the publisher assumes any responsibility for the use or misuse of the information contained in this book. The author and publisher specifically disclaim any personal liability, loss, or risk incurred as a consequence of the use and application, either directly or indirectly, of any advice or information presented herein.

Be advised that there may be certain items represented in this book as to which the **sale, possession, construction** or **interstate transportation** thereof may be **restricted, prohibited** or **subject to special licensing requirements. Consult** with your local **ATF** and law enforcement authorities in your area **before obtaining or constructing such items.**

Technical data and safety procedures presented herein, inevitably reflects the author's individual beliefs and experience with particular equipment and environments under specific circumstances which the reader cannot duplicate or experience exactly. The information in this book should therefore be used for guidance only and should be approached with great caution.

TABLE OF CONTENTS

INGRAM M-10 AND M-11 CONVERSION

The Ingram M-10 and M-11 submachine guns came on the firearms scene in the late 60's and are of the 3rd generation type. In a 3rd generation submachine gun, the front portion of the high mass bolt telescopes around the barrel, thus enabling the weapon to be extremely compact, highly concealable and very reliable. The short distance the bolt needs to travel accounts for the high rate of fire these weapons are capable of. The Ingrams have threaded barrels and were basically designed to be used with sound-suppressors.

Unfortunately, the Ingram submachine guns came at a time when interest in submachine guns was disappearing. The military establishments adopted shorter versions of their assault rifles to replace the submachine gun. Law enforcement agencies, citing poor accuracy and high rate of fire of the MAC's, opted for larger, more accurate weapons such as the HK MP-5 and the Uzi.

The open-bolt semi-automatic versions of the M-10 and M-11 were introduced in the 70's and enjoyed brisk sales until the Bureau of Alcohol, Tobacco and Firearms (BATF) stepped in and prohibited their sales in 1982. The main reason given for this action was the ease of converting these weapons to selective-fire.

At this time only the sale of a modified closed-bolt version of the original design is allowed.

NOTE: The following conversion methods will only work on the original OPEN-BOLT versions of the M-10 and M-11. DO NOT attempt the conversion of a closed-bolt weapon using these instructions.

There are several ways to achieve conversion of the MAC 10-11 to full-auto. The easiest—and least effective—is to grind or file the trip lever (see diagram, part No. 43) until it is completely out of engagement with the bolt. This prevents the trip arm from tripping and disengaging the disconnector from the sear. This is considered a permanent conversion as you would have to replace the complete sear pin bracket in order to return the weapon to semi-automatic only capability.

Another method is to grind or file the left side of the bolt until it completely clears the trip lever. Here again, the trip arm does not disengage the disconnector from the sear. With this method the trip lever does not need to be filed, enabling you to return the weapon to semi-automatic-only capability by merely replacing the modified bolt.

The advantages of these conversions are that you won't have to buy any parts and that they can be done in minutes without special tools. The disadvantage is that the weapon will have full auto capability only. This is not critical in a submachine gun and particularly in the Ingrams as they are specialized, high rate of fire, short-range weapons.

The best method to accomplish the conversion (this is also a permanent conversion) is described in the following pages. For this conversion you will need the following parts.

1. SELECTIVE FIRE SEAR PIN (part No. 37)

2. RETAINER (part No. 50)

3. SEAR STUD (see arrow)

With this conversion the weapon will have SELECTIVE-FIRE capabilities. Your best chance to find the parts needed is at local gun shows or look through old ads in the SHOTGUN NEWS, but hurry as they are getting difficult to find.

The Ingram semi-auto, open-bolt pistols are identical to their submachine gun counterparts, except for the sear pin. To prevent the selective-fire sear from being substituted for the semi-auto version, no sear pin hole has been drilled. Instead, a special sear pin bracket has been spot-welded into the lower receiver and this supports the semi-auto sear pin. To effect this conversion the bracket must be removed and the sear pin hole drilled.

First, grasp the pistol in a firing attitude and make a noticeable mark on the left-hand side of the receiver with a felt-tip pen. Remove the upper receiver and locate the sear pin mounting bracket. You will note that the sear stud passes through the bracket and anchors it to the bottom of the receiver. Removal of the sear stud is the first step of the conversion.

Turn the receiver over so that the grip frame points up and the bottom of the receiver is level. Using the diagram, locate the position for the sear stud hole and mark with a punch indentation. Starting with a smaller bit size and working up to the required diameter, drill a .250" (.187" for M-11) hole through the bottom of the receiver and into the sear stud. (A drill press would be desirable for this operation, but not essential.) This hole does not have to pass through the sear stud, just deep enough to penetrate the receiver wall.

Now, use a thin bladed chisel and attempt to pry the sear pin bracket out of the receiver. If the bracket is attached by other points of weld, carefully drive the chisel between the bracket and the receiver and try to "walk" the bracket up and out.

It is possible, probable in fact, that the sear pin and sear will have to be removed before the bracket can be extracted. Cut the sear pin with a sharp hacksaw and avoid damaging the sear and other parts of the receiver. Remove all the internal components and again try to pry the bracket from the stripped frame. Be careful not to "mutilate" your receiver. (If you do, don't panic. Take it to a friendly gunsmith or machinist—or do it yourself—and have the gouges and scratches filled in with weld. Then file and sand the weld down to the original contours.)

The next step, drilling the sear hole, is the most critical. Place the receiver on its side and locate with a punch mark where the sear pin hole should be drilled. (See diagram for placement.) Make this placement as accurately as possible. Insert a snug fitting block of wood inside the receiver and position it under the drill press. As before, make the first pass with a drill bit smaller than .312" (.250" for M-11) diameter required. After you have the sear pin hole passing through both sides of the receiver, enlarge the hole on the left side to .343" diameter (.281" for M-11). The diagram may confuse you at this point. But remember, you have a felt-tip mark on this side. Insert the new sear into the hole so that the selector switch is on the left-hand side. It should slide in without force, yet have little or no play. Now remove the sear pin.

The last step is to spot-weld the new sear stud into the

.250" (.187" for M-11) hole you drilled earlier. Weld from the bottom of the receiver and then file and sand to original appearances.

To restore a like-new finish to the receiver, apply one of the parkerizing solutions commercially available.

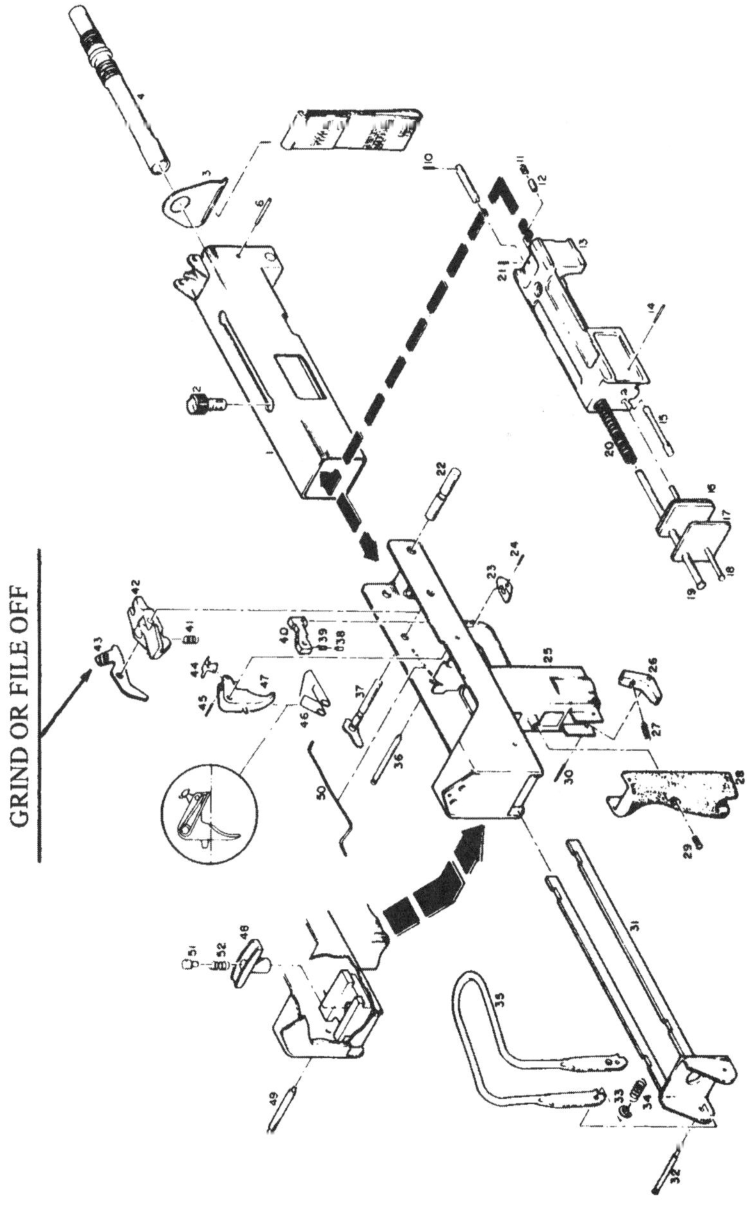
GRIND OR FILE OFF

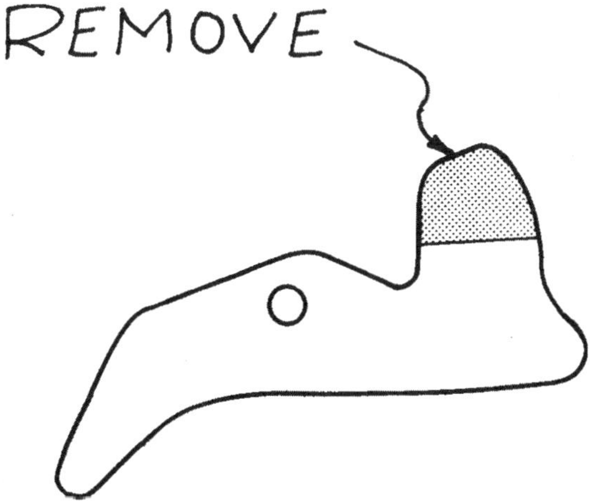

TRIP LEVER #43

NOT TO SCALE

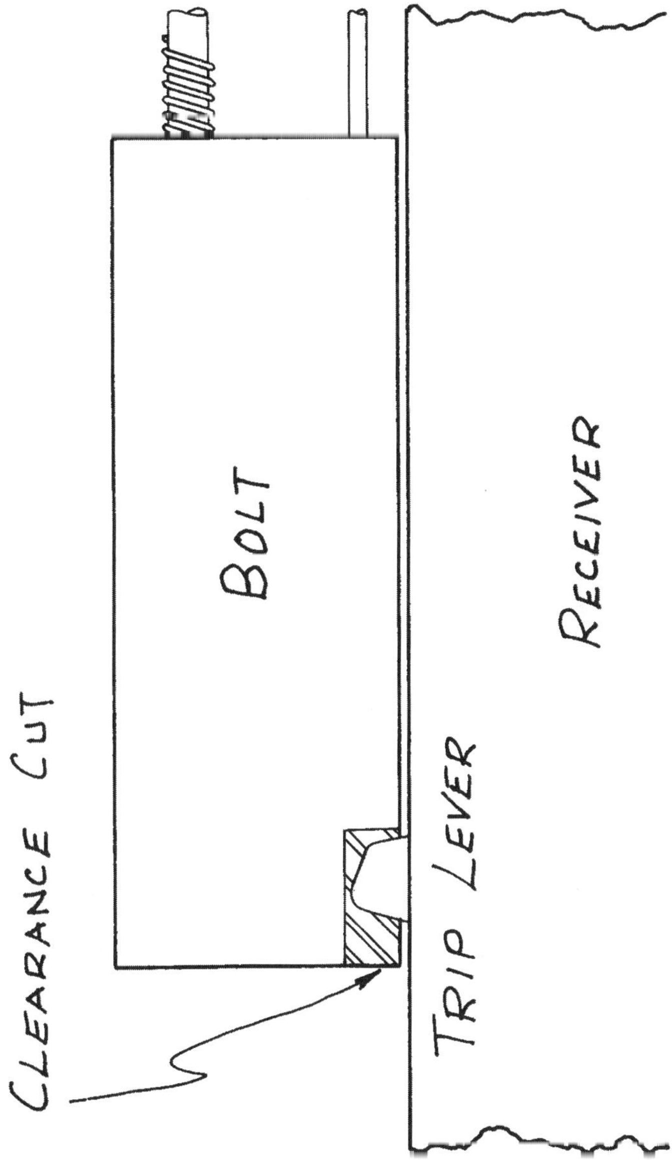
BOLT
RECEIVER
CLEARANCE CUT
TRIP LEVER

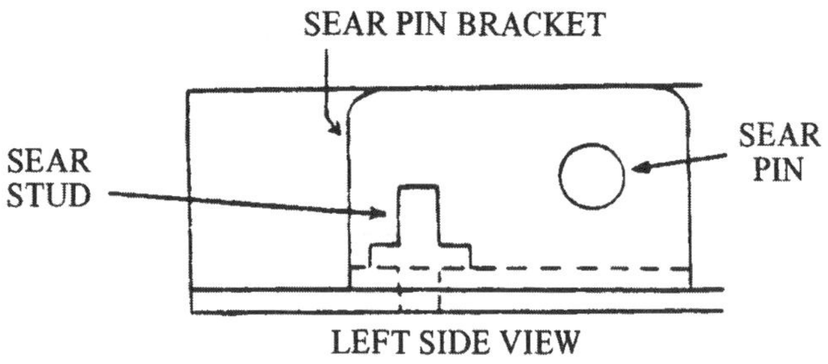
SEAR PIN BRACKET
SEAR STUD
SEAR PIN
LEFT SIDE VIEW

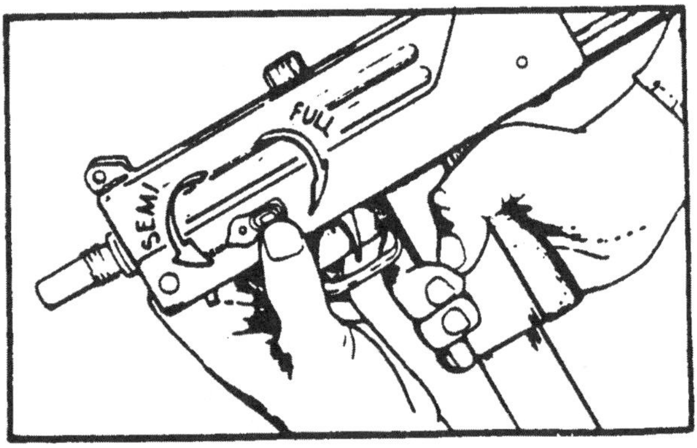
FULL
SEMI

MAC–10 RECEIVER (MAC–11 DIMENSIONS IN PARENTHESES)

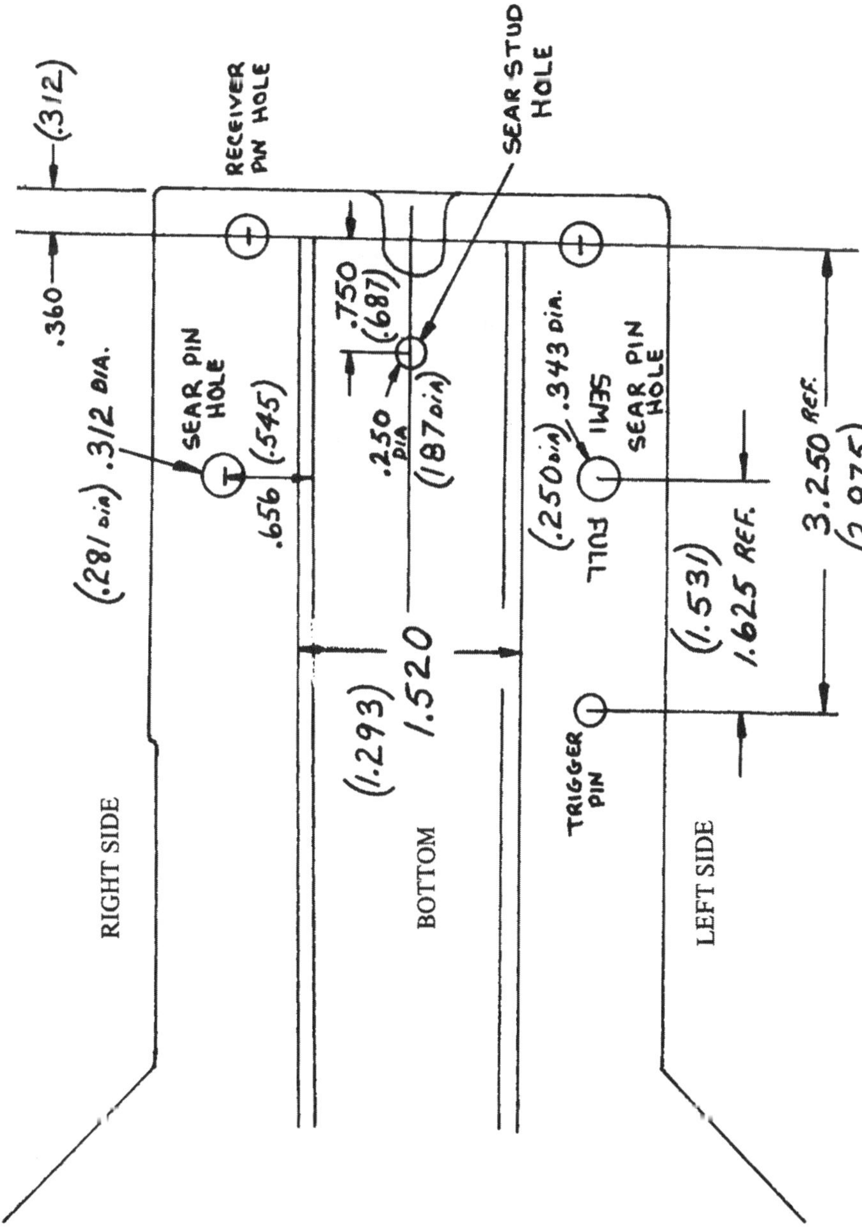

MINI-14 CONVERSION

Introduced to the American public in the 70's by the company of Sturm-Ruger, the Mini-14 has become by far the most popular and best-selling semi-automatic .223 rifle on the market. Its low price, light weight and handy size make it the choice of many shooters.

The Mini-14 rifle resembles the Garand and M-14 military rifles in appearance and mechanical functioning. It is a gas-operated weapon fed by 5, 20 and 30 round magazines.

Full-auto conversion of this weapon is not difficult to achieve, although some machining is necessary. For this conversion, we have used a system similar to the M-14's to select mode of fire. This system requires the addition of one part (Trip Rod) for full-automatic fire. This unit is much easier to build and more reliable than others currently on the market which use a selector lever. Although the selector lever method is much faster and more comfortable for choosing mode of fire, we felt that the two main features of any conversion method should be ease of manufacture and functional reliability; that, we feel, we have accomplished.

To complete the conversion procedure, the following parts need to be made (check drawings).

1. ROCKER

2. ROCKER PIVOT PIN

3. TRIP ARM

4. SECONDARY SEAR PIN

Remove the trigger group and action from the rifle, then remove the operating rod. (Consult your owner's manual where needed.) Drill the necessary holes in the receiver and secondary sear. (See drawings A and B.) You will need to drill a 1/8" hole, .425" from the top of the secondary sear. It is advisable to remove the secondary sear before drilling the hole in this piece. The hole in the receiver should be .157" in diameter and 3/8" deep. Located at .280" from the top of the receiver and .100" from the front.

At this point, make a clearance cut of .062" on the bottom of the operating rod. This is done best using a grinding machine or vertical milling machine. This clearance creates a striking surface and makes the installation of the rocker arm possible. The clearance should begin .062" measured from the front of the receiver when the operating rod is in battery.

Reassembly may now begin. Press the rocker pivot pin

through the rocker and into the hole you have drilled into the receiver. The pre-bent and threaded end of the trip rod must be screwed into the hole on the edge of the rocker. The trip rod, when installed in the weapon, extends past the side of the trigger group and fits between the steel reinforcing plate (in the stock) and the wood. (Some inletting of the stock will probably be necessary.) Due to model and individual weapon variations, some bending of the trip rod may be necessary in the area of the trigger group.

Once the trip rod extends back to the hole in the secondary sear, a 90 degree bend must be made in the rod so that it extends through the hole drilled in the secondary sear. (IMPORTANT: Be sure that the length of the trip rod from rocker to secondary sear is such that the bolt is locked before the secondary sear is tripped!) When you have the trip rod through the hole in the secondary sear, secure it in place by installing a pin through the end of the rod, screwing on a lock washer, or similar means of preventing the rod from vibrating out through the hole. DO NOT MODIFY THE SECONDARY SEAR OTHER THAN DRILLING THE HOLE DESCRIBED ABOVE!

With this unit installed, the weapon will fire full-auto only with a cyclic rate approximating 750 rpm. It can be quickly converted back to semi-auto by removing the trip rod.

For a slightly different version of this modification, see drawings B and C and instructions below.

Trip arm must be inletted into the wood on the right hand side behind metal reinforcement and must be trimmed to length at final assembly to insure bolt is locked before secondary sear is tripped. Trip arm works between rocker and press fit pin in the secondary sear.

The only parts different with this method are the trip arm, secondary sear (needs pin) and the rocker (no trip arm hole is necessary). Everything else remains the same, including functioning and rate of fire.

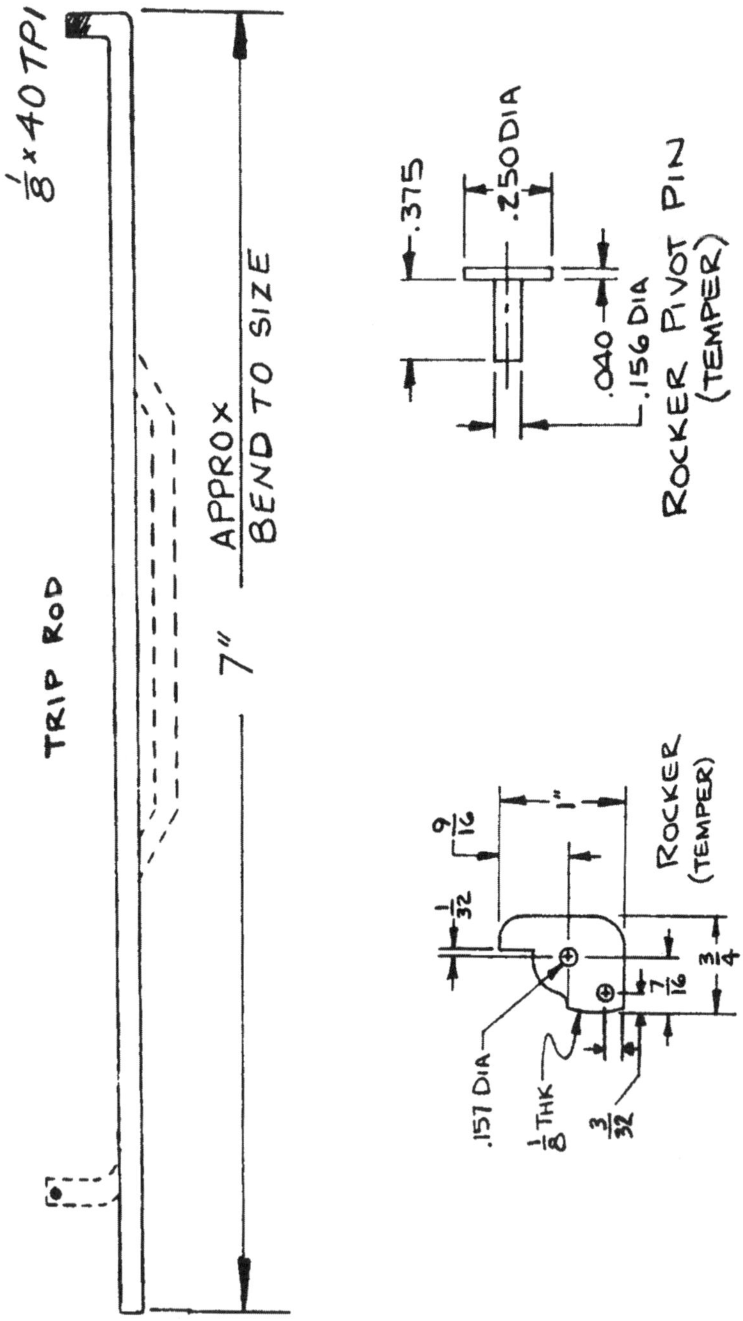
1/8 x 40 TPI
TRIP ROD
APPROX
BEND TO SIZE
7"
.250 DIA
.375
.040
.156 DIA
ROCKER PIVOT PIN
(TEMPER)
ROCKER
(TEMPER)
1"
9/16
1/32
3/4
7/16
.157 DIA
1/8 THK
3/32

DRAWING A

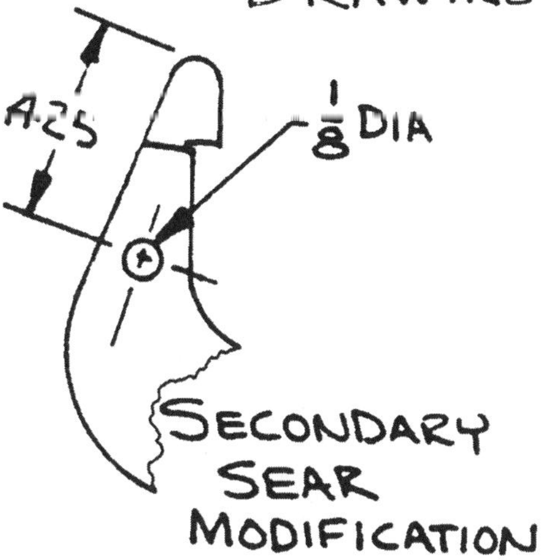

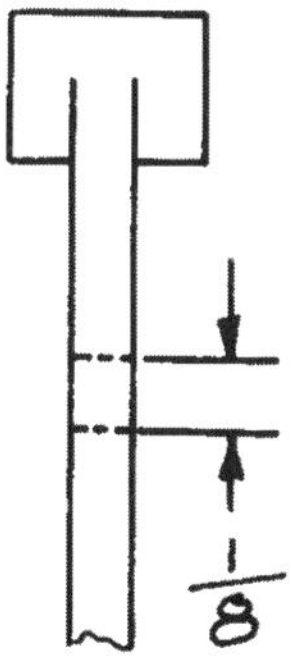

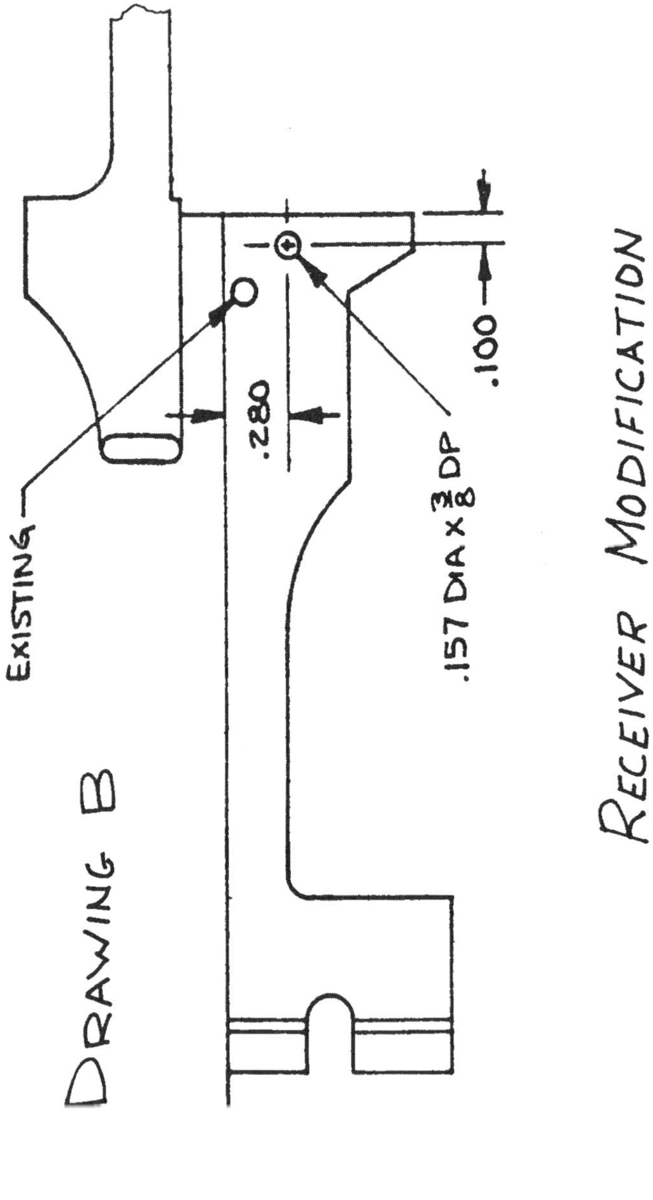

Drawing B
Existing
.280
.157 DIA x 3/8 DP
.100
Receiver Modification

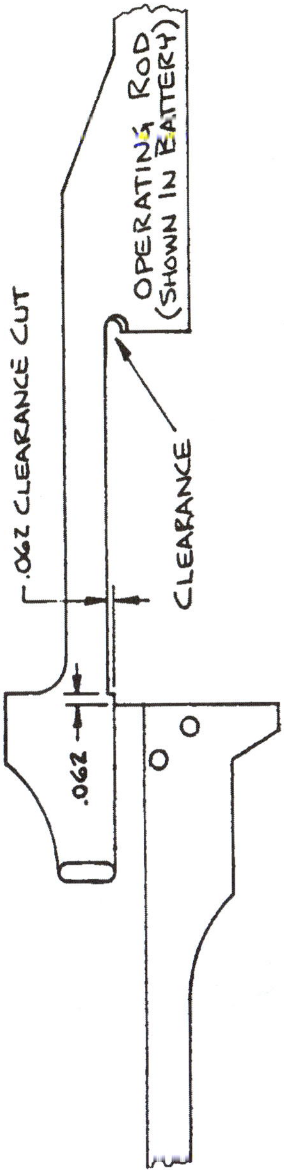

OPERATING ROD MODIFICATION

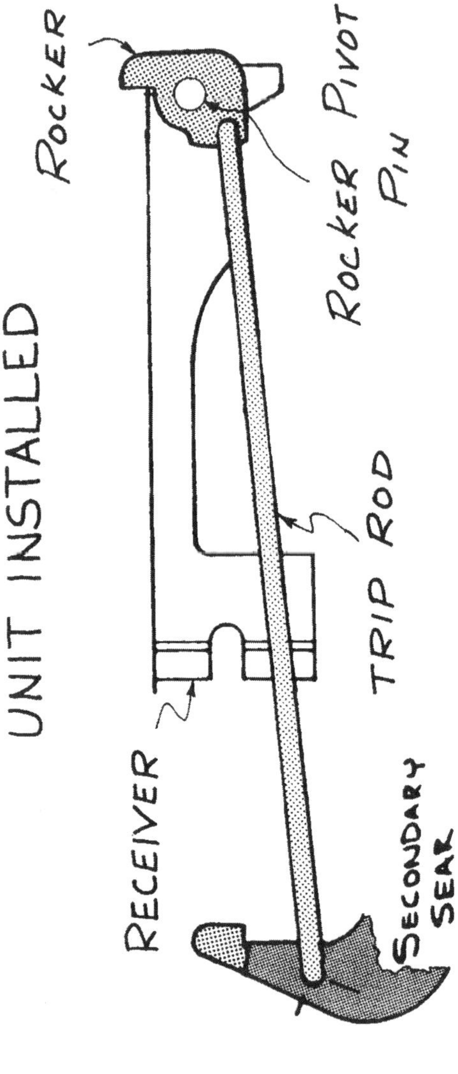
UNIT INSTALLED
ROCKER
ROCKER PIVOT PIN
TRIP ROD
RECEIVER
SECONDARY SEAR

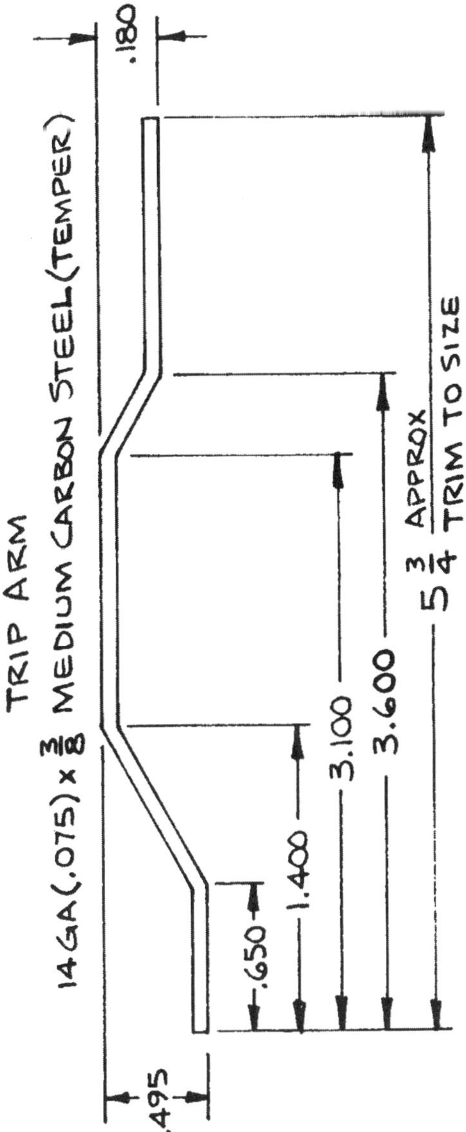
TRIP ARM
14 GA (.075) x 3/8 MEDIUM CARBON STEEL (TEMPER)
.180
.495
.650
1.400
3.100
3.600
5 3/4 APPROX
TRIM TO SIZE

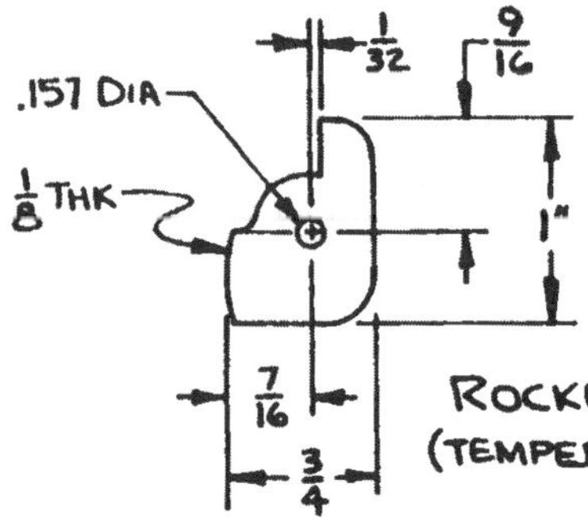

ROCKER
(TEMPER)

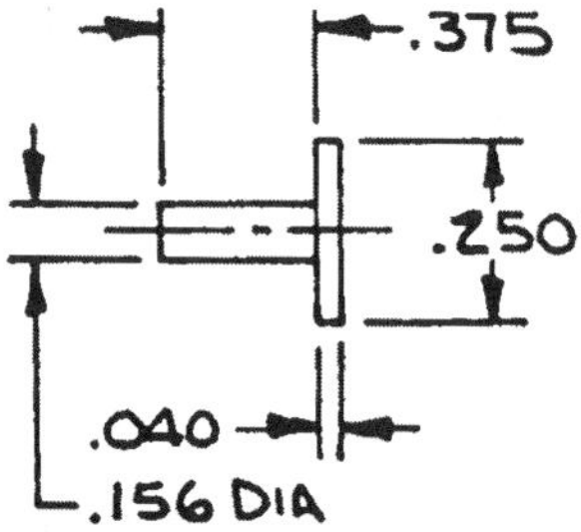

ROCKER PIVOT PIN
(TEMPER)

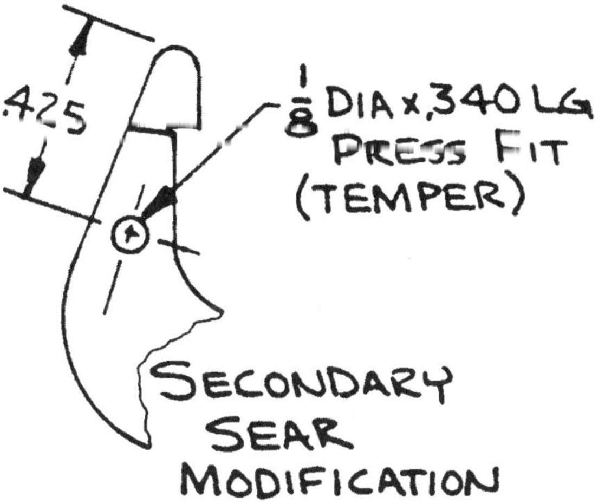
.425
1/8 DIA x .340 LG
PRESS FIT
(TEMPER)
SECONDARY
SEAR
MODIFICATION

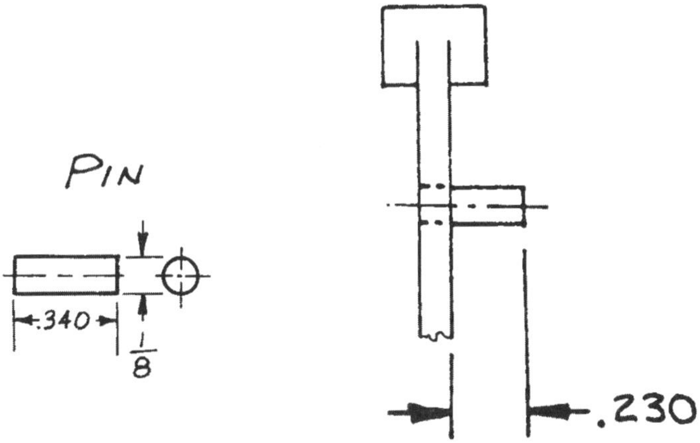
PIN
.340
1/8
.230

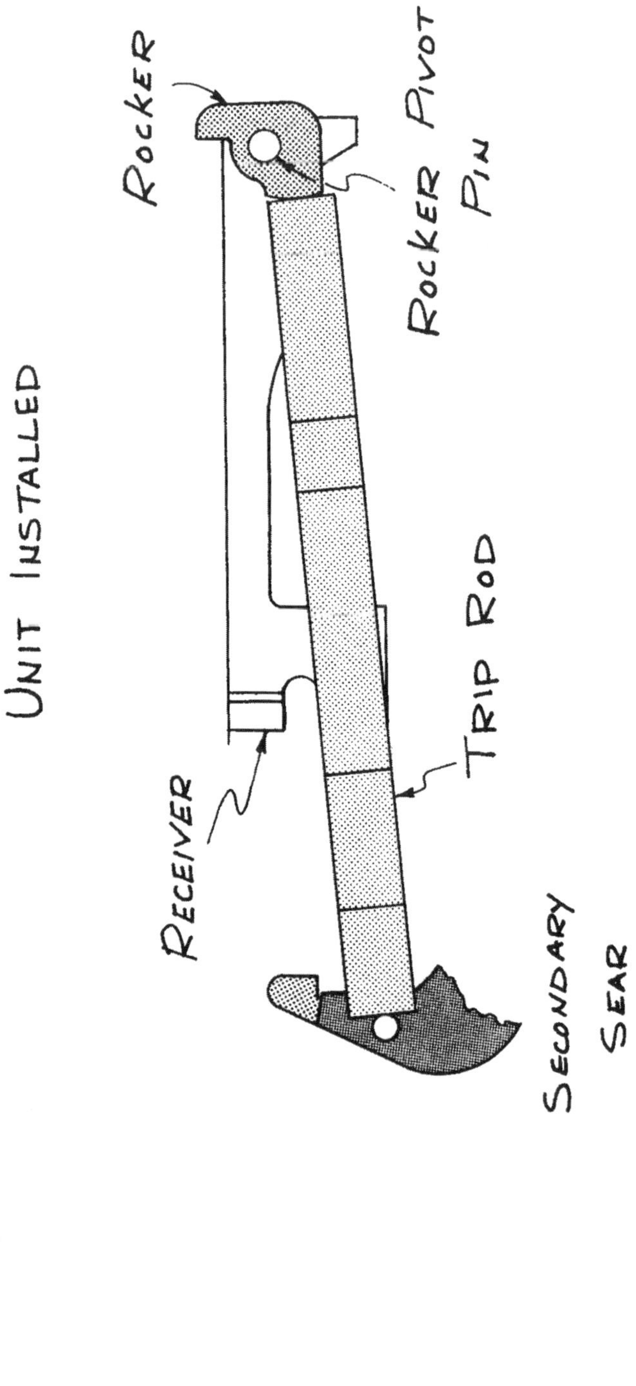
UNIT INSTALLED
ROCKER
ROCKER PIVOT PIN
TRIP ROD
RECEIVER
SECONDARY SEAR

H & K 91/93 CONVERSION

The HK 91-93 rifles imported and sold commercially in the U.S. are semi-automatic versions of the standard German service rifle known as the G-3.

The G-3 is a development of the Spanish CETME assault rifle, which in turn, was developed from a German World War II rifle design. The semi-automatic versions are almost identical to their selective-fire counterparts; except for the exclusion of the automatic mechanism and the different specifications of several parts.

To successfully accomplish the select-fire conversion of the Heckler and Koch weapons follow the directions that follow —carefully—taking the time necessary to do a good professional job. Be sure to use the best materials available as indicated and remember: These conversions are illegal if you do not have a Federal license. DO NOT TAKE ANY CHANCES WITH THE LAW! If you are licensed to do the work, then proceed as follows.

Remove the buttstock and pistol grip assembly; use your owner's manual as a reference. Rotate the safety selector to the straight-up position and pull the selector out. Now lift the complete Trigger Group Housing (No. 62) from the pistol grip housing.

Examine the Trigger Group Housing (TGH) and compare it to the illustration in Figure A. Hold it so that it is pointing away from you with the hammer forward and uncocked. Place a pointed instrument in the yoke of the hammer spring retaining rod (No. 75) and pull backwards and up away from the hammer. Remove the rod and spring. Now, hold the TGH with the left side up and the trigger pointing towards you. Use a blunt instrument to lift the trigger spring (No. 68) over the retaining ledge of the trigger and out of the TGH. Remove the hammer pin (No. 79) and hammer (No. 74); trigger pin (No. 73), trigger (No. 67) and sear (No. 72). Lastly, remove the sear rollerspring pin (No. 66) and the sear roller spring assembly (No. 63 and No. 64).

With the TGH dissassembled, cut or file (a hacksaw works best) a notch on the bottom of the housing (Figure B). This notch creates room for the sear roller spring elbow to pivot into when the connector lever is depressed. If you cut the slot deeper than .175", don't panic. This affects nothing. In fact, if the slot is not deep enough, problems in functioning may result.

The next step is to remove the sear retaining pin (No. 71) and replace it with the shaft of a 5/64" drill bit (Figure F). Be careful when performing this operation as the sear detent piston spring (No. 69) is under pressure and may escape. Note: Make the drill bit long enough to just fit inside the

TGH, but with enough clearance to move freely up and down inside the housing.

A detent for the full-auto position must be made on the pistol grip housing (Figure E). Place a snug fitting hardwood block inside this housing. Using a center punch, ground to fit the existing safety detent, form a full-auto detent midway between the Safe and Fire positions.

Now, examine the stock trigger and compare it to Figure D. You will note the selector notch has to be enlarged and shaped to the angle shown. You may find later that a width of .185" may not be sufficient to allow the trigger to trip the hammer. If so, enlarge the notch to the semi-circle shown.

The side of the trigger in the shaded area has to be honed. This side fits snugly against the connector lever. However, there should be enough clearance for the connector lever to move freely.

Begin reassembly by placing the connector lever (No. 78) inside the TGH. Replace the sear roller spring and pin (No's. 63, 64 & 66). Note: One leg of the elbow spring should align with the retainer groove in the connector lever. Assemble the sear, trigger and spring as one unit. Place in the TGH and install the trigger pin.

Pivot the sear roller spring (No. 64) over the sear, bringing forward pressure on the trigger.

Return the trigger spring (No. 68) back into the retaining ledge of the trigger.

Replace the hammer (No. 74), hammer pin (No. 79), hammer spring (No. 76), and the retainer rod (No. 75).

The last phase of the conversion, modifying the selector lever, requires a certain amount of hand-fitting and experimentation.

Place the assembled TGH inside the pistol grip housing. Insert the selector into the pistol grip housing, cock the hammer and put the safety on "F".

Follow the next procedure carefully. It duplicates the rifle's movements when performing automatic fire.

Press down on the connector lever and hold, then squeeze the trigger until the hammer falls. Keeping the trigger back, release the connector lever, then push the hammer back until it catches. Press the connector lever down. The hammer should fly forward if all of your modifications have been correct. At this point your rifle in firing full-auto at the "F" (or semi) position. Now repeat this procedure (which we will refer to as the firing procedure) with the selector at the middle detent. Hopefully, the results will be the same. If not, you will have to enlarge the selector notch on the trigger. (See optional semi-circular notch on Figure D.) After you have modified the trigger properly, go on to the next step.

Using Figure C as a reference and a file, carefully file the selector to the required shape. Periodically, return the selector to the pistol grip and place in the "F" position. Repeat the firing procedure described in the above paragraph. You have filed enough when the connector lever is pushed down the second time and the hammer does not release. Now, try this with the safety on the middle (Auto) detent. The hammer should still release when the connector lever is pushed down the second time. Note: Be careful not to file too deeply on the selector or the rifle will only fire semi-auto on either detent.

Put the whole weapon together and test fire. You will be pleased to note that our conversion allows for a 100-150 rpm lower cyclic rate than the original G-3. Consequently, the weapon is more controllable in full-auto mode.

59. PISTOL GRIP TRIGGER ASSEMBLY

60. SAFETY SELECTOR

62. TRIGGER GROUP HOUSING (TGH)

63 & 64. SEAR ROLLER SPRING ASSEMBLY

66. SEAR ROLLER SPRING

67. TRIGGER

68. TRIGGER SPRING

69. SEAR DETENT PISTON SPRING

71. SEAR RETAINING PIN

72. SEAR

73. TRIGGER PIN

74. HAMMER

75. HAMMER SPRING RETAINER ROD

76. HAMMER SPRING

78. CONNECTOR LEVER (CONVERSION LEVER)

79. HAMMER PIN

81. EJECTOR

DRAWING A

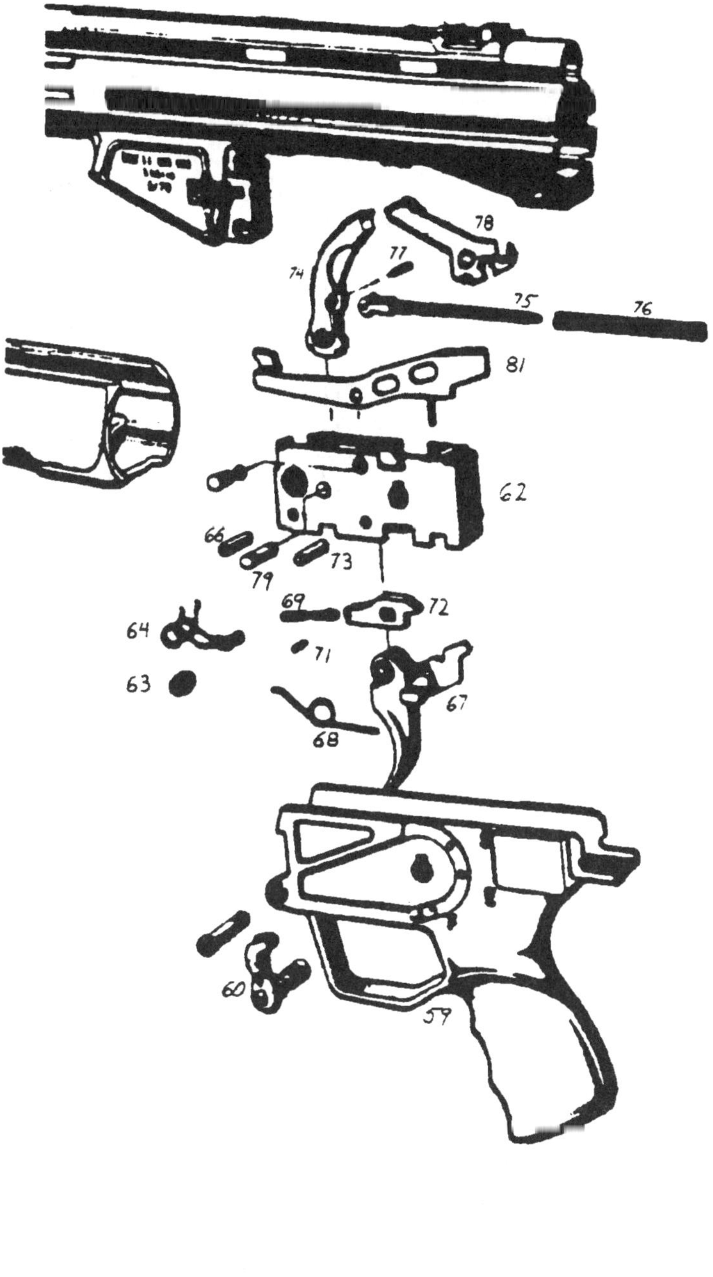

DRAWING B

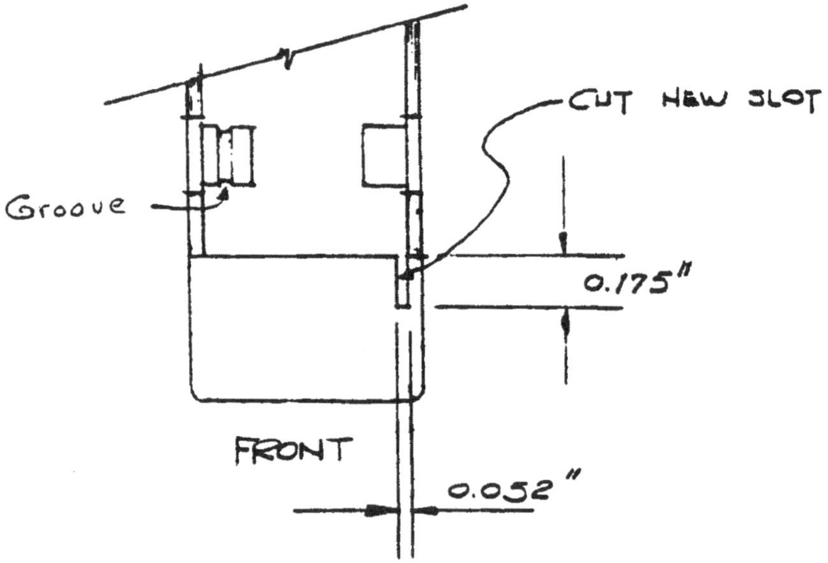

BOTTOM VIEW
TRIGGER GROUP HOUSING #62
NOT TO SCALE

DRAWING C

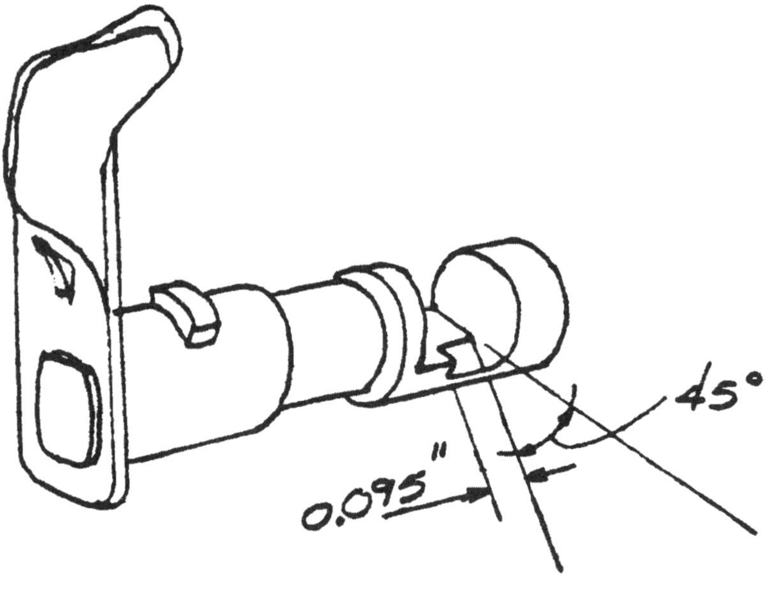

SAFETY SELECTOR #60

DRAWING D

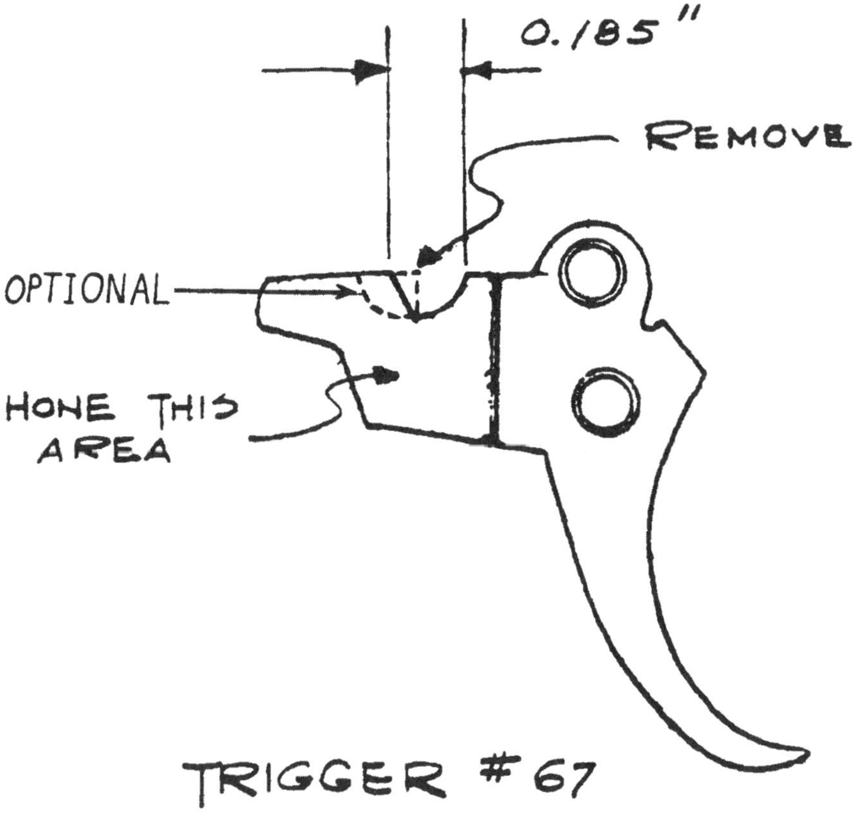

TRIGGER #67
NOT TO SCALE

DRAWING E

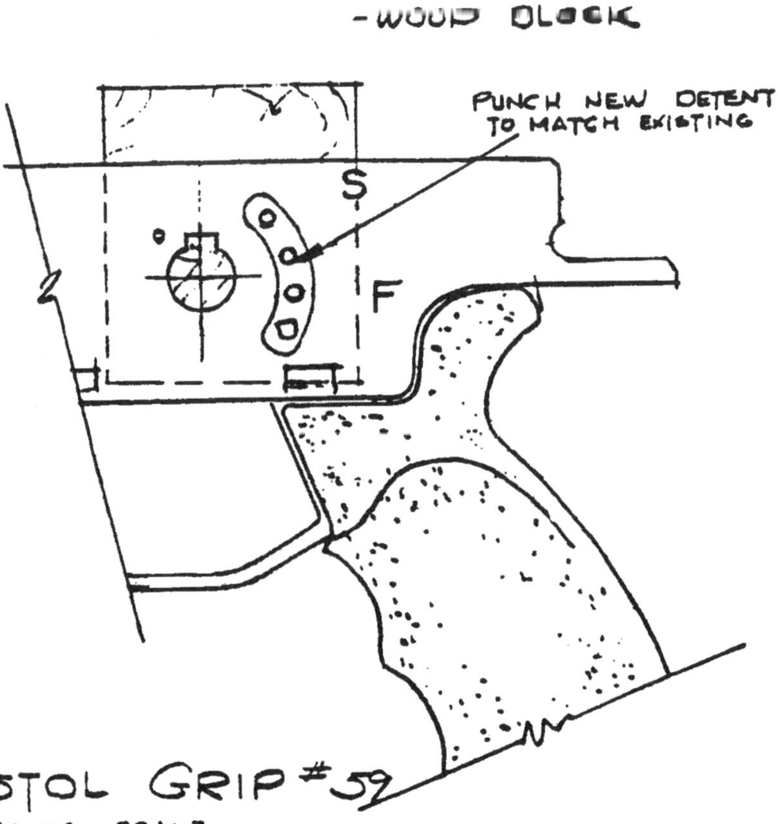

PISTOL GRIP #59
NOT TO SCALE

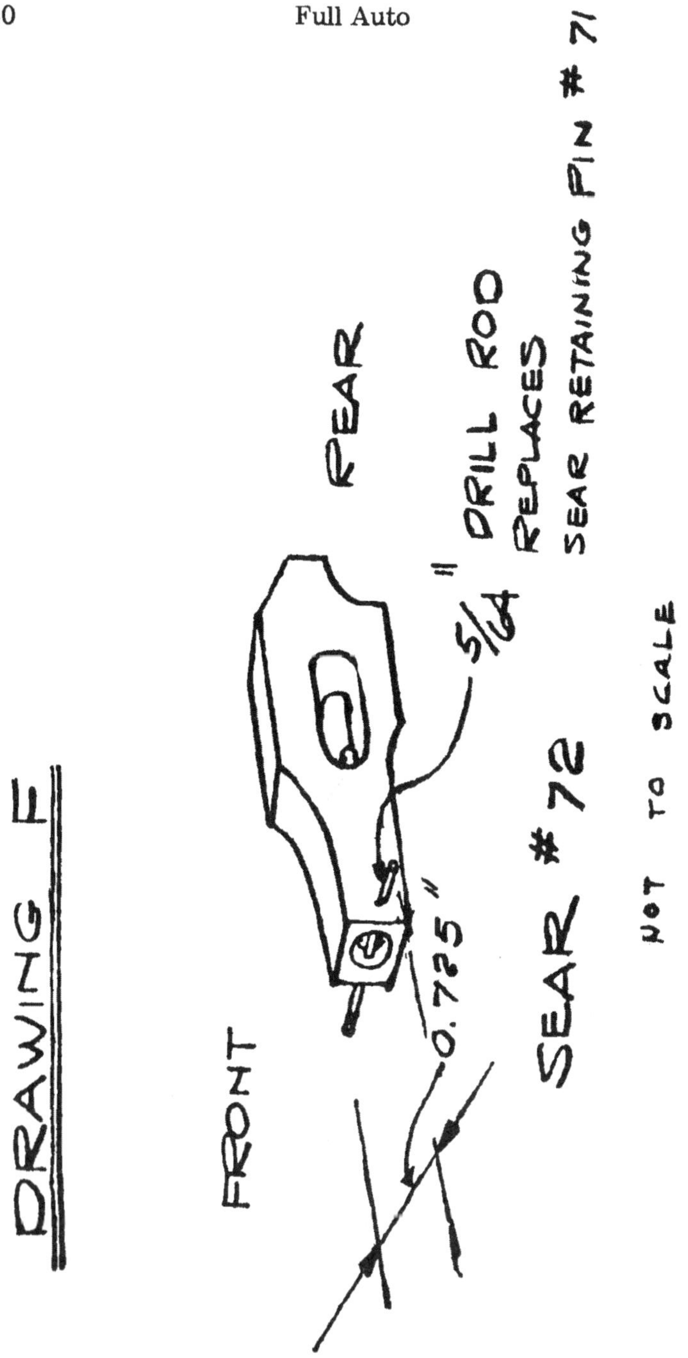
DRAWING F
FRONT
REAR
5/64" DRILL ROD
REPLACES
SEAR RETAINING PIN # 71
0.725"
SEAR # 72
NOT TO SCALE

HK 91/93 CONNECTOR LEVER BLUEPRINTS

MATERIAL:

A 2" x 4" piece of 15 guage (.0673") sheet steel.

TOOLS:

14.5 mm Metric drill (.570").
Drill press or hand drill.
Bandsaw or jigsaw.
File.
Vise.
Hammer.
Scriber or sharp nail.

PROCEDURE:

Cut out template drawing or make one from scratch paper by tracing around the drawing. This drawing is done to scale and is the same size as the actual part. Apply Dyken or another anti-reflective agent on the 15 guage piece of sheet steel and let dry. This will allow you to see the tracing steel, and use a scriber or sharp nail to trace around it. Mark the center where the pivot hole must be drilled.

Using a drill press or hand drill, drill the pivot hole, starting with a smaller bit size and working up to the 14.5 mm (.570") required.

After you have drilled the pivot hole, check the tracing again to make sure the hole is in the proper position. If not, use Dykem to block out the old trace, properly align the template drawing with the drilled hole and retrace.

Cut around tracing with a bandsaw or jigsaw, getting as close to the tracing as possible, but not going over the lines. Use a file to smooth any rough edges, checking your lever with the template drawing as you go, to insure you are getting the correct shape. If a caliper is available use it to get a more accurate measurement.

At this stage, place the template drawing over the lever and mark where you will be making a 90 degree bend. Place the lever in a vise with the mark parallel to the jaws of the vise, but about .010" or the thickness of a penny, above the top of the jaws. Use a hammer to bend the sheet steel 90 degrees TOWARD THE RIGHT HAND SIDE, WHEN LOOKING AT IT FROM THE REAR! See drawing for a clearer view.

Again, place the template drawing on the lever and mark the area you will need to hone to about half the thickness of .034". This is best done on a grinding machine or milling machine, but if none are available, use a file, starting with a coarse grade and finishing with a fine toothed file to leave a smooth finish; emery cloth can also be used.

A 3/4" long, .030" deep groove must be made in the lower front edge of the connector lever. This groove is tricky and it is recommended you use a Dremel tool, with a thin abrasive wheel to grind it while holding the lever in a vise. This groove should be about 1/32" wide.

At this point you are ready for heat treating. The best way is to take the lever to a professional steel treating shop and have them case harden it to a depth of .010" or .015". If you wish to do it yourself, use Kasenit or similar case hardener available at many hardware stores. Follow instructions on the can.

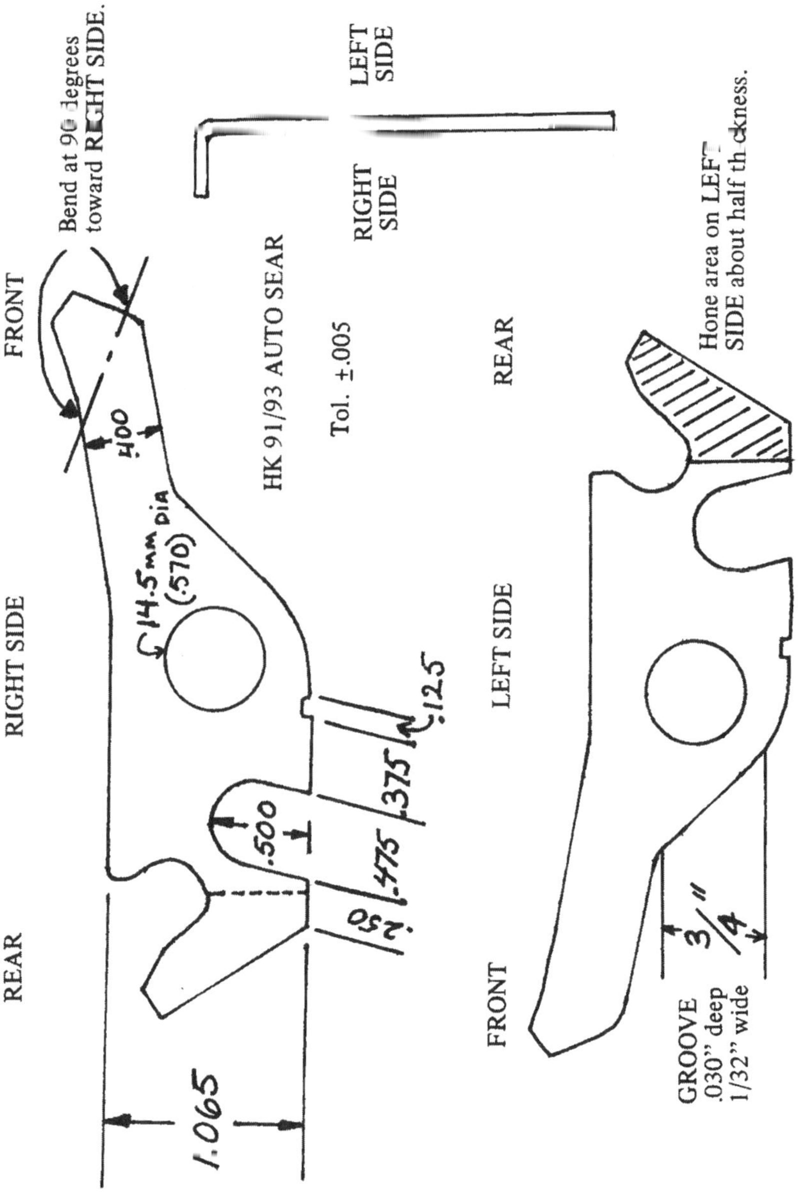

Bend at 90 degrees toward RIGHT SIDE.
FRONT
REAR
RIGHT SIDE
.400
14.5MM DIA (.570)
.500
.250
.475
.375
.125
1.065
HK 91/93 AUTO SEAR
Tol. ±.005
RIGHT SIDE
LEFT SIDE
REAR
LEFT SIDE
FRONT
Hone area on LEFT SIDE about half thickness.
3/4"
GROOVE .030" deep 1/32" wide

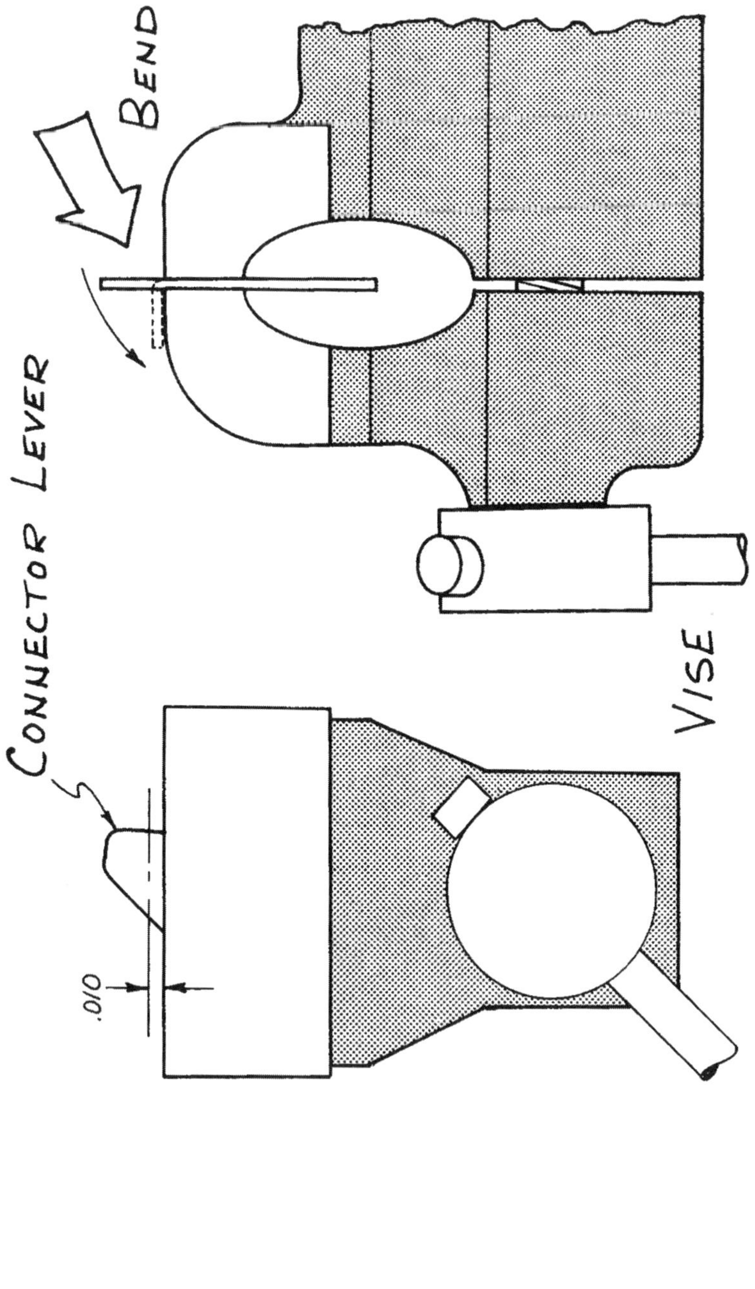
BEND
CONNECTOR LEVER
VISE
.010

AR-15 CONVERSION

Selective-fire conversion of the Colt AR-15 is a simpler operation than with most other weapons. All that's required is the interchanging of a few "civilian" AR-15 parts for the corresponding "military" M-16 parts, and the addition of a single part.

The M-16 has been the official service rifle of the U.S. Armed Forces since 1966, so parts are easily obtained and for the most part, inexpensive. This rifle has gone through several design changes during the past 20 years and is now finally considered a reliable weapon.

To convert this weapon to selective-fire capability, you will need the following M-16 parts.

1. BOLT CARRIER (use your AR-15 bolt)

2. DISCONNECTOR

3. HAMMER

4. TRIGGER

5. SELECTOR

In addition, you will need an auto sear (Figure 6). This part can be taken out of the weapon in seconds to return it to semi-auto only capability. There is no machining needed to install the auto sear.

To install the M-16 parts, follow the instructions below:

Field strip the rifle using your owner's manual as a reference.

Carefully remove the pistol grip from the lower receiver by removing the screw and lock washer. Catch the helical spring and selector detent to prevent them from getting lost.

Remove the hammer and spring by pushing out the pin from the receiver. The hammer should be forward with the selector in the fire position.

Next, remove the trigger and disconnector. Using a drive pin punch, push out the trigger pin from the left side of the receiver. Be careful not to lose the disconnector helical spring.

With the trigger and disconnector out, you will be able to remove the selector lever.

To complete the conversion, reassemble the lower receiver replacing the AR-15 parts with the M-16 parts.

Install the M-16 selector lever in the receiver. Set in the semi-auto (arrow pointed up) or full-auto (arrow pointing to the rear) position.

Remove the AR-15 trigger spring and disconnector spring and install the M-16 trigger. Insert the disconnector in the slot on the trigger, while aligning the holes. Place the trigger and disconnector in the receiver and insert the pin from the right side of the receiver.

Next, install the AR-15 hammer spring in the M-16 hammer and insert hammer into the receiver. Be sure that the ends of the hammer spring are installed to the rear of the hammer and trigger pin. The ends of the hammer spring must be resting in the annular grooves on the upper surface of the trigger pin.

Insert the selector detent (pointed end up) and spring from the bottom of the receiver. Carefully compress the spring with the pistol grip and secure the grip in place with the lock washer and screw.

Finally, remove the bolt, firing pin, firing pin retaining pin and bolt cam pin from the AR-15 bolt carrier and install in the M-16 bolt carrier. Refer to your owner's manual as the AR-15 parts assemble in the M-16 bolt carrier in exactly the same manner.

The final step in this conversion is the installation of the auto sear at the rear of the lower receiver. (See photo for placement.)

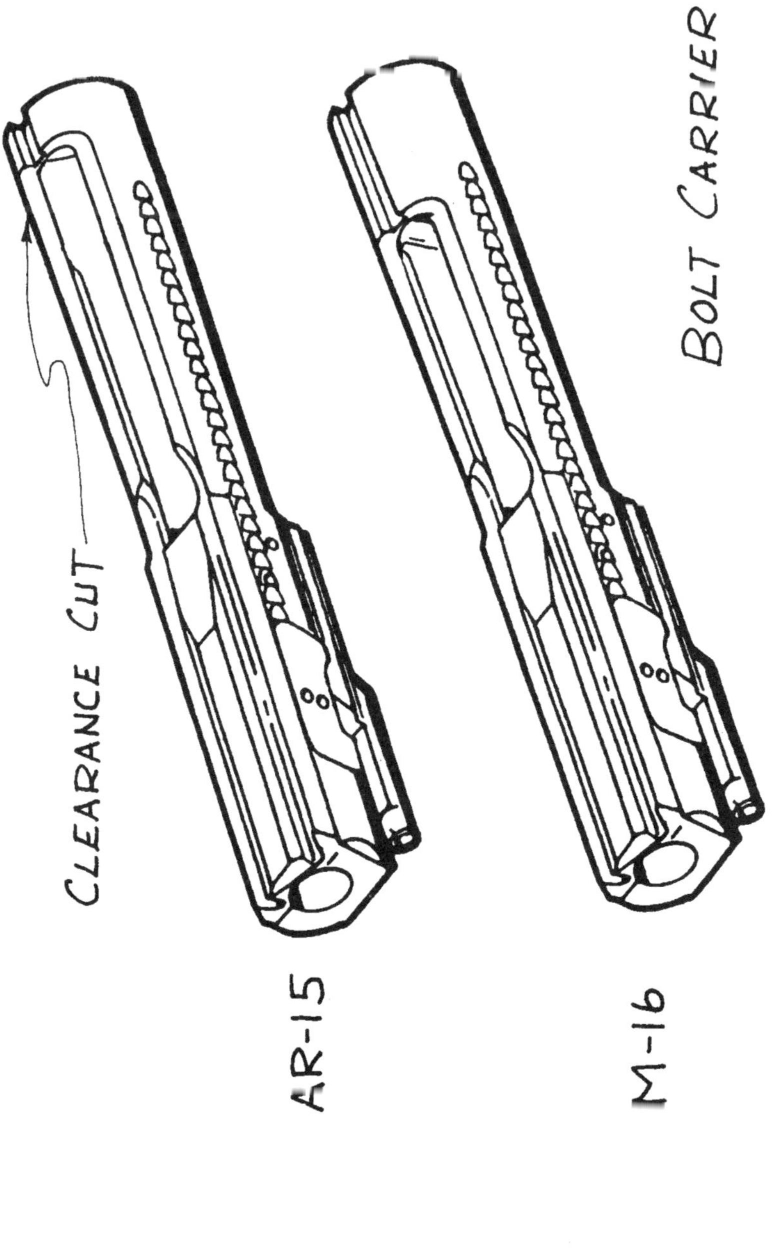
CLEARANCE CUT
AR-15
M-16
BOLT CARRIER

DISCONNECTOR

AR-15

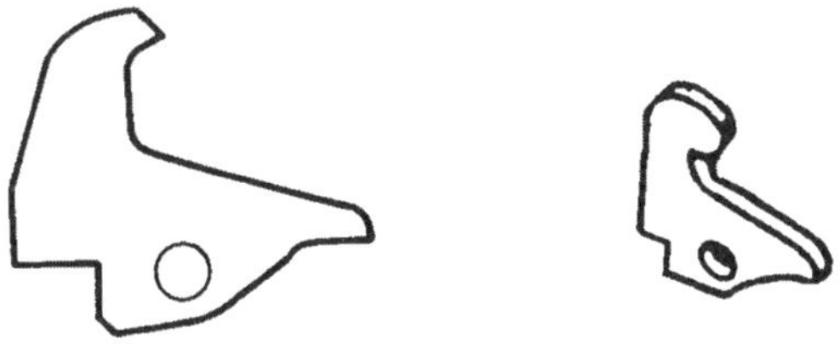

M-16

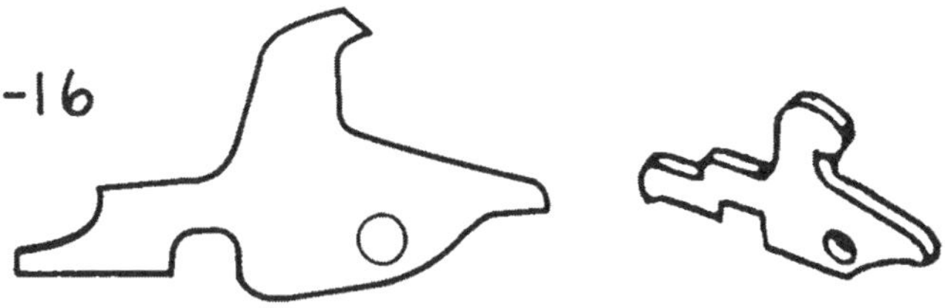

HAMMER

AR-15

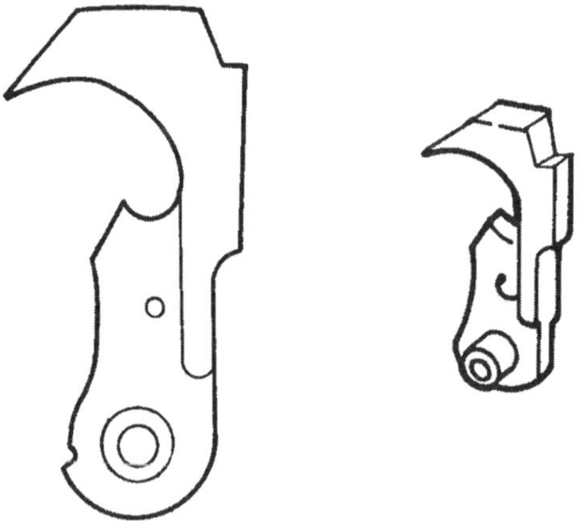

M-16

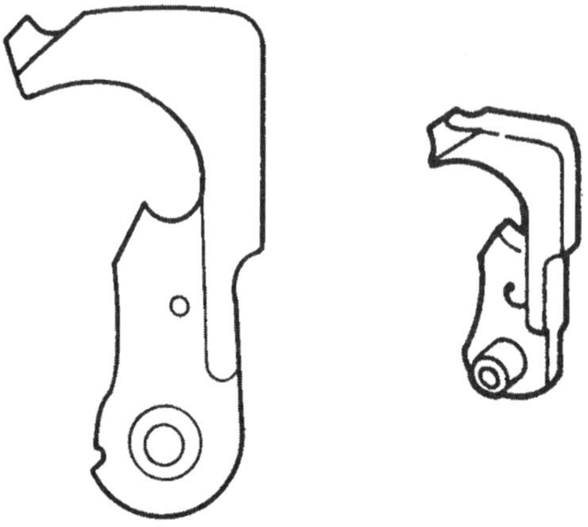

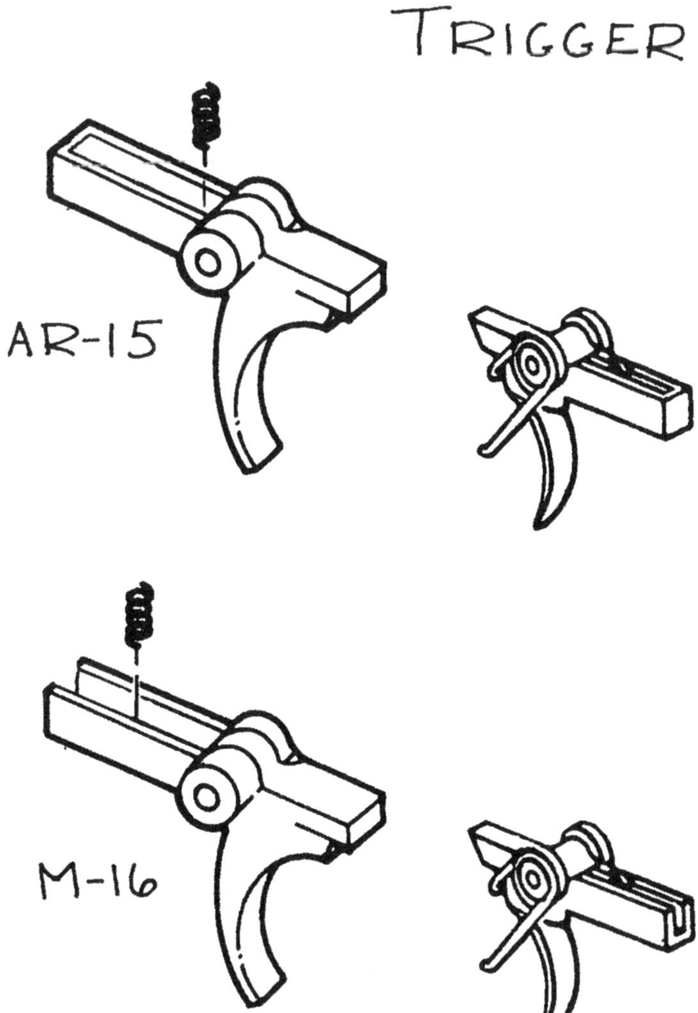
TRIGGER
AR-15
M-16

SELECTOR

AR-15

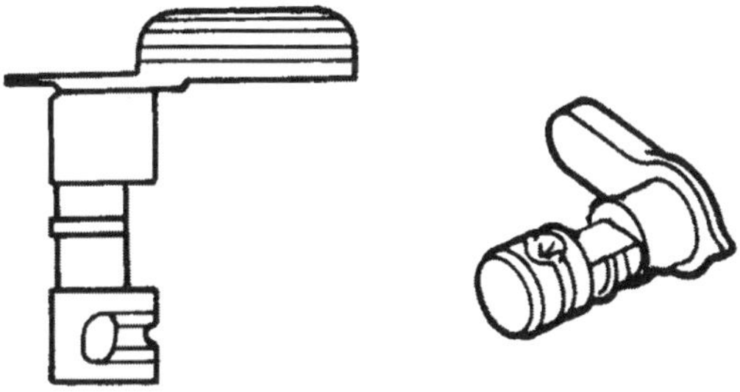

M-16

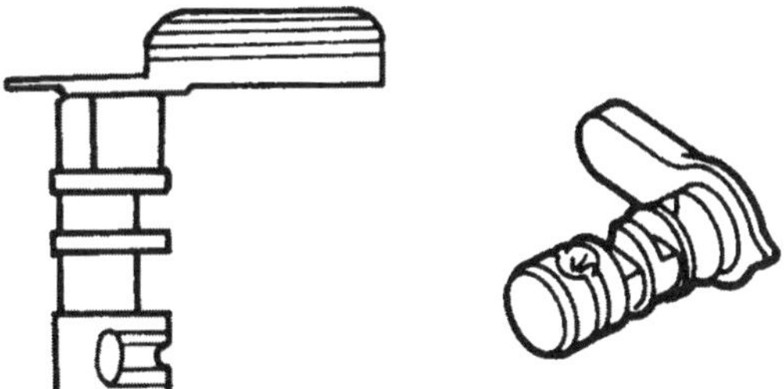

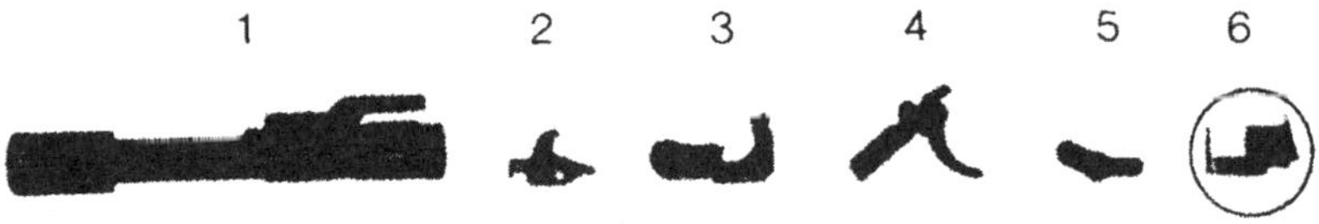

1. BOLT CARRIER (use your AR-15 bolt)
2. DISCONNECTOR
3. HAMMER
4. TRIGGER
5. SELECTOR

AUTO SEAR (Figure 6)

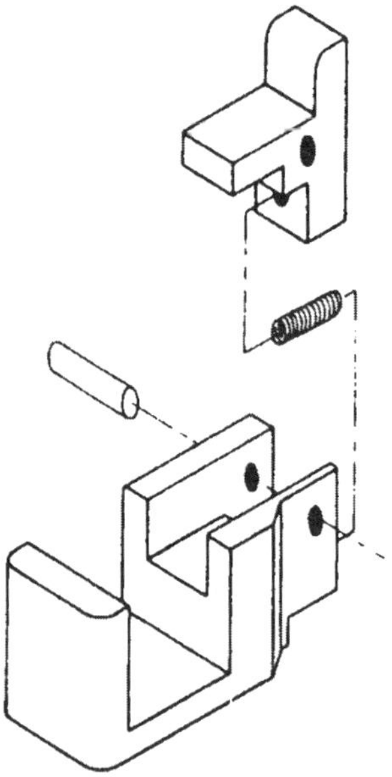

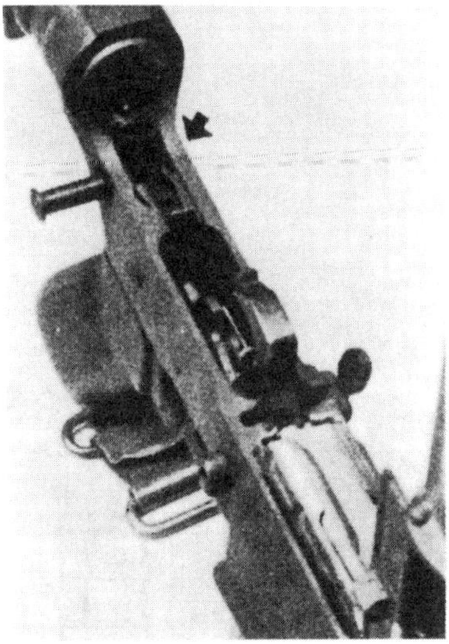

TO INSTALL THE AUTO SEAR JUST "DROP IT" IN AS SHOWN IN PICTURE.

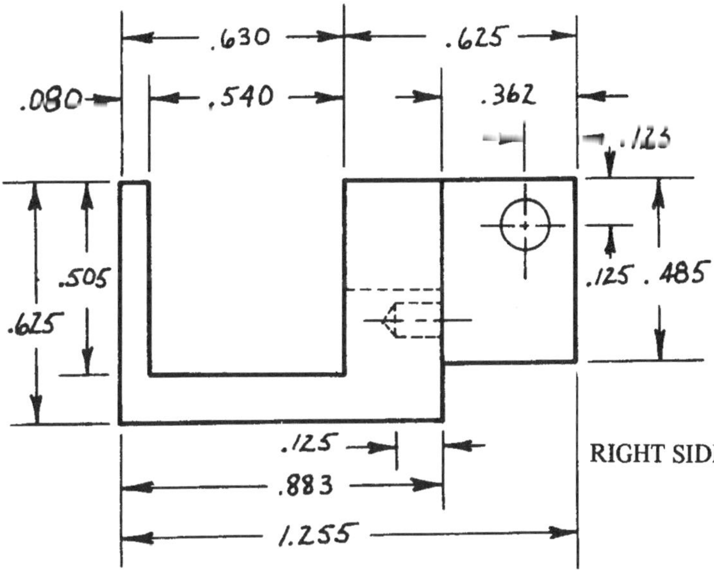

RIGHT SIDE

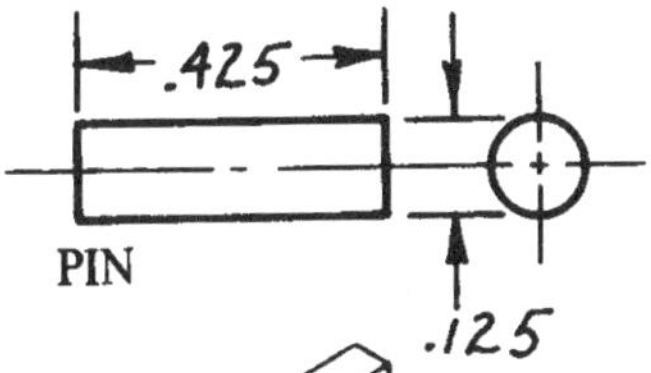

PIN

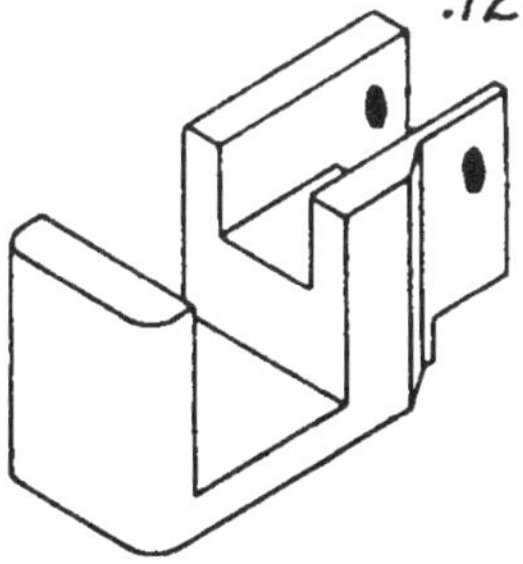

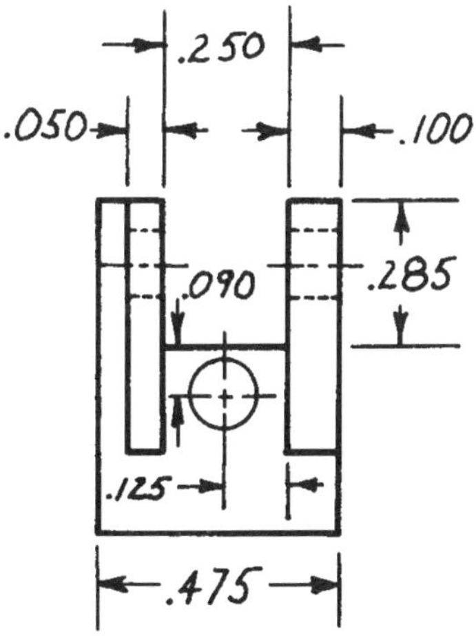

AR-15 AUTO SEAR BODY

ALL HOLES 1/8" DIA. Tol. ±.005

FRONT

TRIP ARM

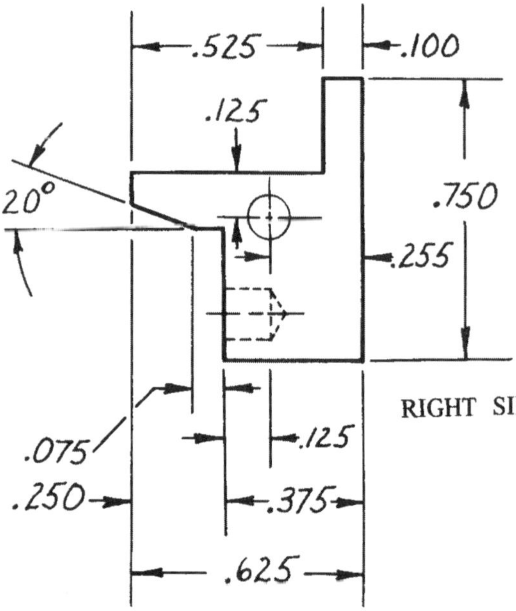

RIGHT SIDE,

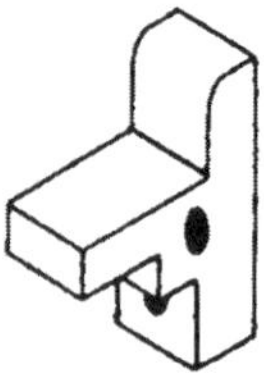

BACK

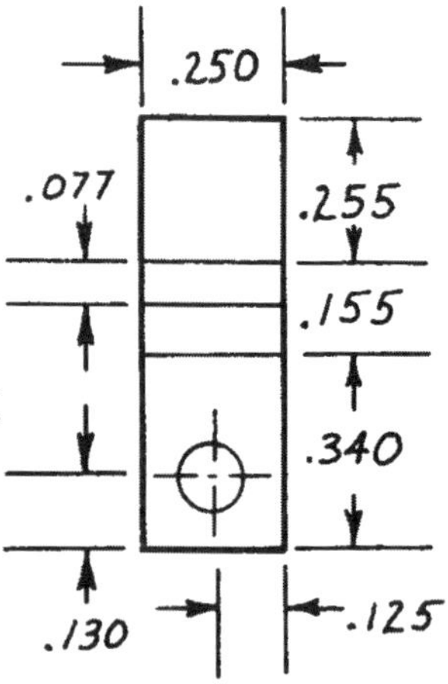

The trip arm should be manufactured from quality tool steel or 4140 steel. Use drill rod for the pivot pin. All parts should be tempered, except the spring.

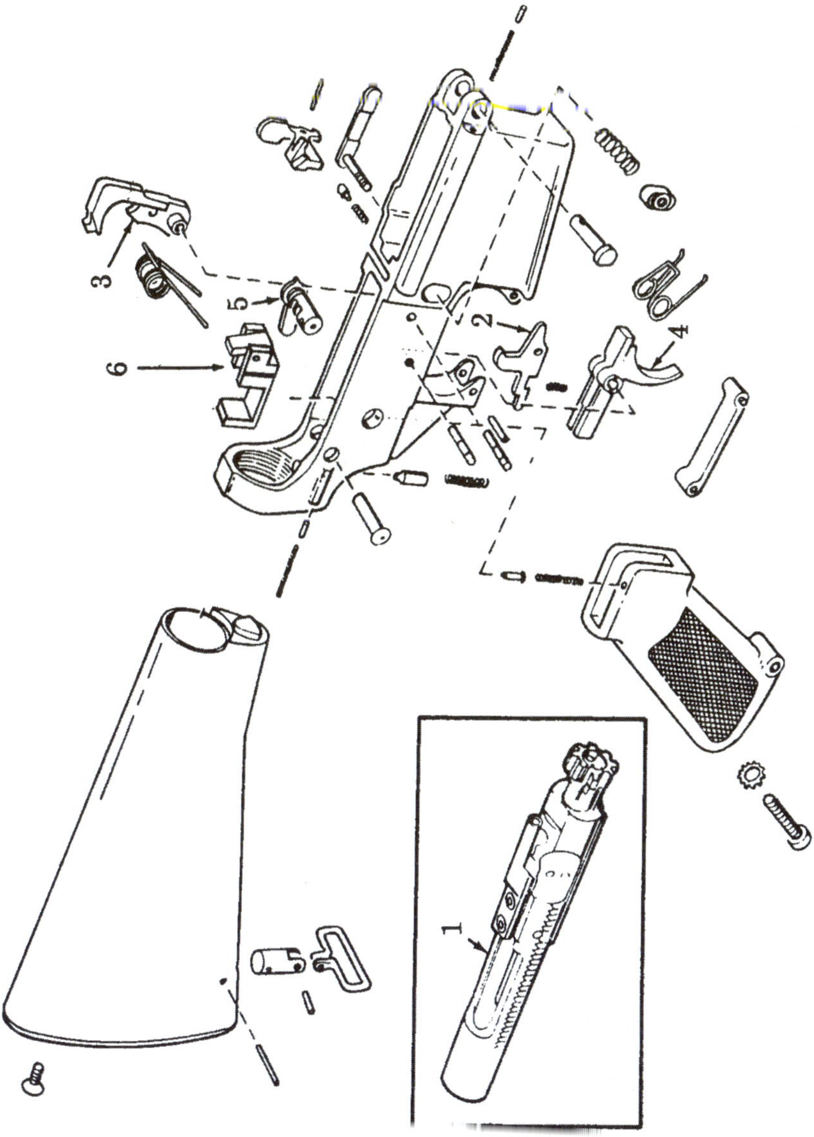
3
5
6
2
4
1

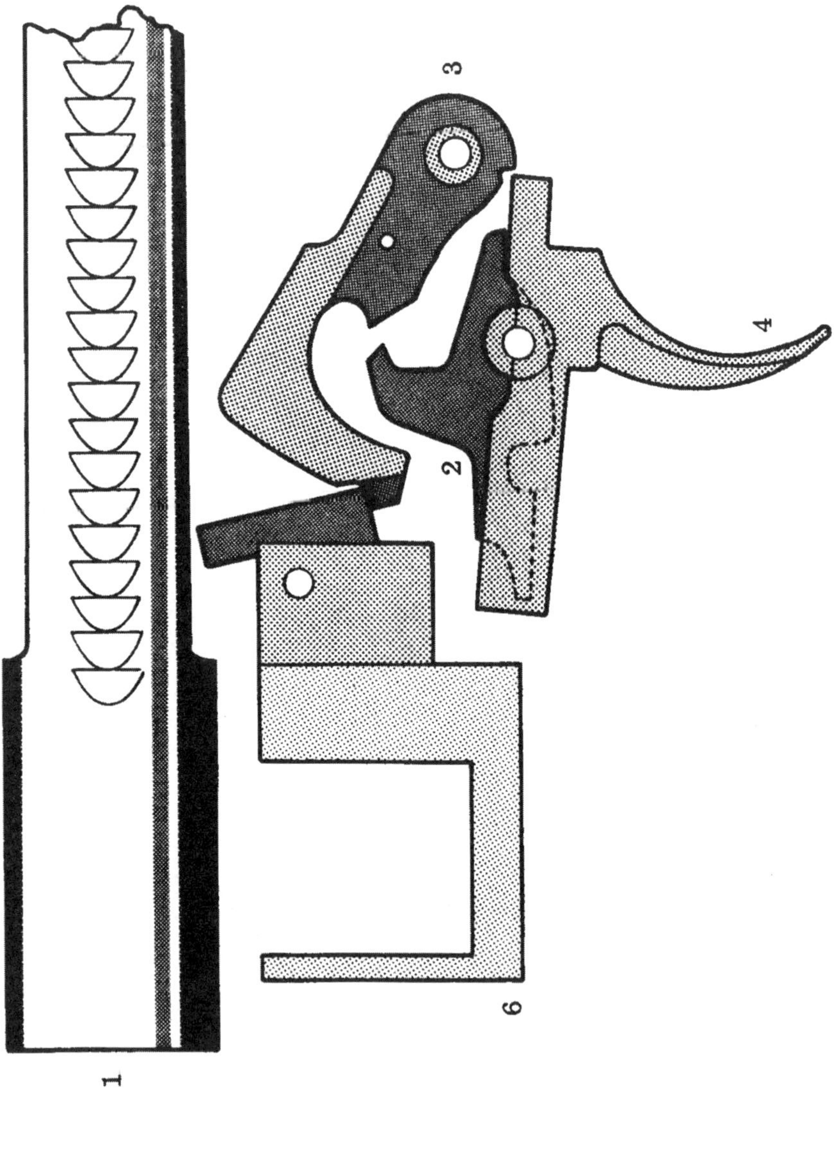
3
4
2
6
1

M-1 CARBINE CONVERSION

NOTE: Due to design changes, the conversion parts mentioned below WILL NOT FIT the Universal M-1 Carbine.

The M-1 .30 caliber Carbine is a gas-operated, self-loading, air-cooled shoulder weapon. It is usually fed either by 15 or 30 round magazines. The M-2 Carbine is the selective-fire version of this same weapon and ownership is illegal.

The M-2 Carbine is the same as the M-1 except for the differences in design of certain components and the addition of others which permit the M-2 to deliver either semi-automatic or full-automatic fire.

Component differences between full-automatic and semi-automatic Carbines include:

1. HAMMER: The hammer of the M-2 Carbine is the same as the M-1 Carbine, except that it has a milled cut in the lower right side to furnish clearance for the disconnector when assembled on the hammer pin.

2. SEAR: The sear of the M-2 Carbine is the same as the M-1, except for a raised shoulder on the top of the front end, which forms a camming surface for the disconnector when operated.

3. TRIGGER HOUSING: The trigger housing of the M-2 is the same as that of the M-1, except that the left side of the magazine port is furnished with a retention slot and the front face of the port with a dismounting notch for the selector spring. (The dismounting notch leads into the top of the slot.) The right side has a milled cut for clearance of the disconnector lever. The left side has a milled cut for the selector.

4. OPERATING SLIDE: The operating slide of the M-2 is the same as that of the M-1, except for a clearance cut extending along the right-hand side of the body and a diagonal cut at the point where the shank of the handle joins the body. The latter cut forms a cam for camming down the forward end of the disconnector lever.

5. STOCK: A clearance cut is made in the inner right wall of the M-2 stock to provide clearance for the projecting right side of the disconnector. A cut is also made in the inner left wall for clearance for the selector. The bridge is cut down to the central section for clearance for the disconnector lever.

6. DISCONNECTOR GROUP—Added parts: The disconnector pivots on the hammer pin when assembled. The rear end has a lateral projection, which bears upon the raised shoulder of the sear, when the disconnector is cam-operated by the disconnector lever for full automatic fire. The forward end has a projecting lug on the right side, which extends outside the trigger housing and engages and acts as a camming surface for the rear end of the disconnector lever. A spring and a plunger, recessed in the top of the disconnector and

bearing on the receiver, return the disconnector to the inoperative position, when the camming action of the lever is discontinued.

7. DISCONNECTOR LEVER ASSEMBLY—Added parts: The disconnector lever assembly is not to be disassembled. The disconnector lever assembly is composed of a disconnector lever pin, disconnector lever rivet and disconnector lever. The pin retains the trigger housing when assembled to the receiver. The disconnector lever, riveted to a pivot on the pin, is shifted in a vertical plane, by the turning of the pin. The pin is turned by the selector. A rounded projection on the rear end of the disconnector lever operates the disconnector. A projecting toe on the front end of the disconnector lever contacts the camming surface on the operating slide. An offset in the rear section provides for alignment with the disconnector.

8. SELECTOR GROUP—Added Parts: The selector is mounted to the left end of the crank pin by means of a slot in the lower forward face of the selector mating the straddle slots in the end of the pin. The selector holds the pin in position and acts as a lever for turning, throwing the disconnector lever into or out of engagement with the operating slide. A curved wire spring holds the selector in position on the pin and in the full-automatic or semi-automatic position when operated. The straight front end of the spring seats into a recess in the lower rear end of the selector and the circular rear end of the spring seats in a vertical slot in the front face of the magazine post on the left side of the trigger housing. When assembled, the circular end of the spring is downward (spring is concave downward).

As can be seen from the preceding paragraphs, the differences between the "civilian" M-1 and the full-automatic M-2 are minimal. Conversion of the M-1 Carbine to selective-fire is a simple matter and is accomplished by simply interchanging the required M-2 parts in place of the M-1 parts.

You will need the following M-2 parts to succssfully complete your conversion. It is recommended that you use all of the M-2 parts for reliability and safety. A possible exception may be the M-1 stock, which can be modified by cutting or notching enough clearance for the selector and other operating parts.

You will need the following M-2 parts.

1. HAMMER

2. SEAR

3. TRIGGER HOUSING

4. OPERATING SLIDE

5. STOCK (or modified M-1 stock)

6. DISCONNECTOR GROUP—Consisting of:
   a) disconnector
   b) disconnector plunger and spring

7. DISCONNECTOR LEVER ASSEMBLY—Consisting of:
   a) disconnector lever pin
   b) disconnector lever rivet
   c) disconnector lever

8. SELECTOR SWITCH AND SPRING

9. MAGAZINE CATCH (Optional)

10. BOLT (Round type)

NOTE: Although some of the parts in the semi-auto M-1 Carbine could be modified to M-2 specifications, this would require a great amount of work, knowledge of machine shop operations, several formed cutters (which are very expensive) and at the very least a vertical milling machine.

As you can see, it would be very difficult for the average person to complete the conversion (despite what a few ridiculous publishers claim in their ads), using "homemade" parts. We do NOT recommend that you machine or alter any part in the M-1 Carbine except to modify the stock.

To complete this conversion follow the instructions below:

Field strip your M-1 using your owner's manual as a reference.

You will need to completely disassemble the M-1 trigger housing to use some of the parts.

With the hammer forward, insert a pointed instrument in the yoke of the hammer spring retaining rod and pull backwards. Push spring and rod to one side while releasing pressure to remove from the trigger housing. Push hammer pin out of the housing and remove hammer.

Remove the trigger spring. Push the trigger pin out and remove the sear, sear spring and trigger.

To remove the magazine catch, place your thumb over the catch to prevent it from getting lost and insert a pointed instrument into the hole at the bottom of the trigger housing. Depress the magazine catch retainer plunger assembly and remove catch. Remove retainer plunger assembly and safety.

You are now ready to assemble the M-2 trigger housing. Install the safety and retainer plunger assembly. Carefully insert the magazine plunger assembly into the recess hole on the right side of the M-2 trigger housing. Use a pointed instrument to depress the retainer assembly and install the magazine catch.

Install the M-2 trigger and sear (with its spring). Place the sear spring in the recess hole in the sear and insert both into the slot on the trigger. Align all holes with the triggor housing and insert the trigger pin. Install the trigger spring.

To assemble the disconnector and hammer, place the disconnector in trigger housing so that the lug projects over the right side of the housing. Insert the hammer on the left of the disconnector so that its milled lower cut presses against the flat portion of the disconnector. Insert the hammer pin from the left side and through the hole in the hammer, line up the hole in the disconnector and push the pin through the hole in the other side of the housing. With the pin head on the left side, the pin is prevented from shifting when the carbine is assembled. Install hammer spring.

Before you can install the M-2 trigger housing to the receiver, you will have to install the bolt and M-2 operating slide. Refer to your owner's manual as assembly of the bolt and M-2 operating slide is exactly the same as with the M-1.

To install the M-2 trigger housing to the receiver, insert disconnector spring plunger assembly into the hole in the top of the disconnector, spring first. Line up the trigger housing with the receiver and insert the disconnector lever pin from the right side with the long leg of the lever towards the front. The rear toe of the lever should rest on the bottom surface of the projecting lug of the disconnector. Slide the slot of the selector through the grooves of the projecting portion of the disconnector lever pin. Pivot the selector to the rear position and insert straight end of the selector spring into the recess in the lower rear end of the selector; with the loop of the spring downward insert the loop end into the slot in housing, magazine left post mounting slot and down to the bottom of the post.

Finally, install the operating slide guide and spring as with the M-1.

The M-1 stock can be modified for use in this conversion. Make a clearance cut on the right side for the disconnector and disconnector lever and clearance cut on the left side for the selector. Install the modified stock to complete conversion.

NOTE: The field stripping of the M-2 is the same as for the M-1 except for the following:

In order to remove the trigger housing group and the operating slide, the disconnector lever must first be removed. The disconnector lever pin locks the housing to the receiver and prevents the removal of the slide.

To remove the disconnector lever, move the selector to its rearward position and push the selector spring up the slot in the magazine housing post to the dismounting notch and withdraw, using a pointed instrument or the operating slide spring guide. Slide the selector off the disconnector lever pin. The disconnector lever can then be withdrawn and the trigger housing separated from the receiver.

Please be aware that mere possession of all the above-mentioned parts constitutes possession of an illegal automatic weapon, whether assembled or not.

HAMMER

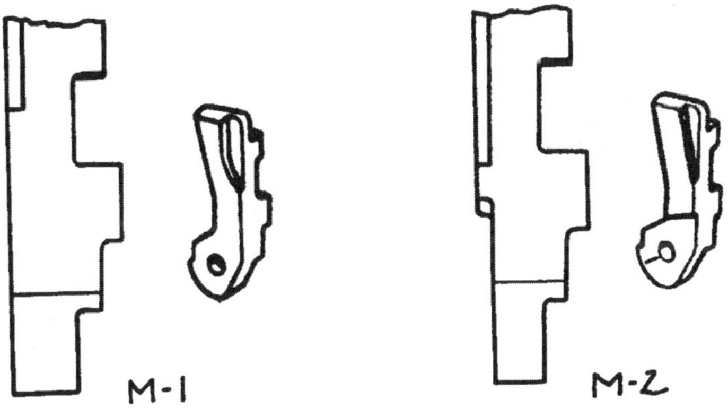

SEAR

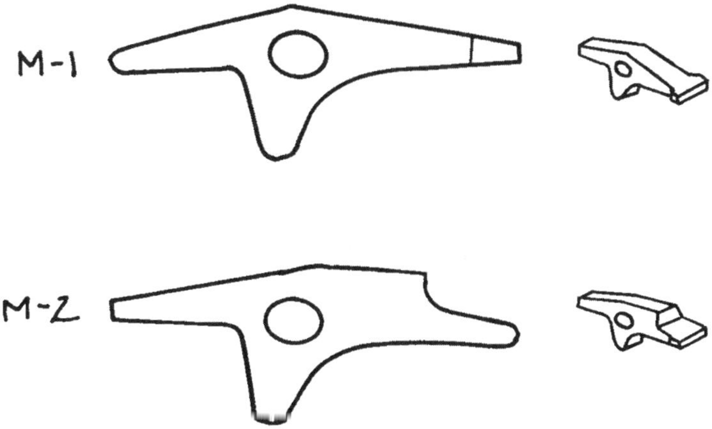

TRIGGER HOUSING

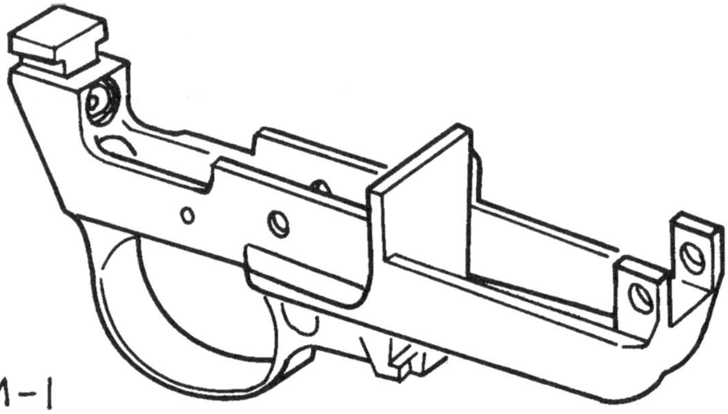

M-1

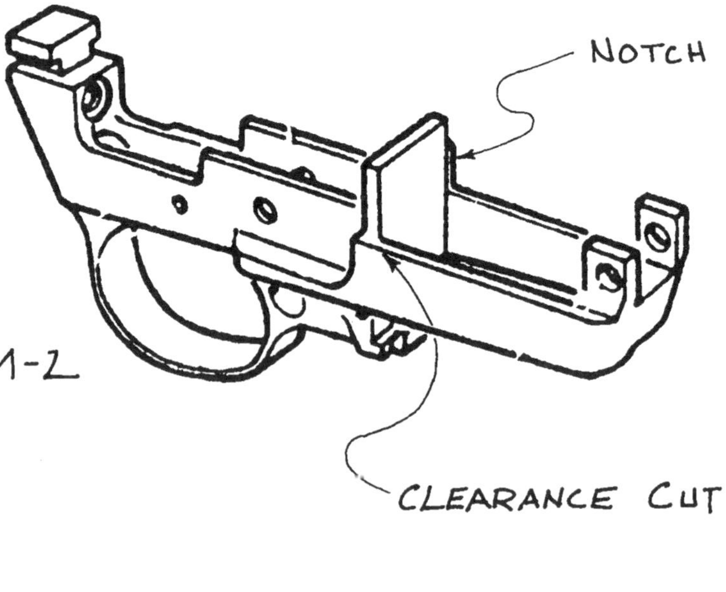

M-2

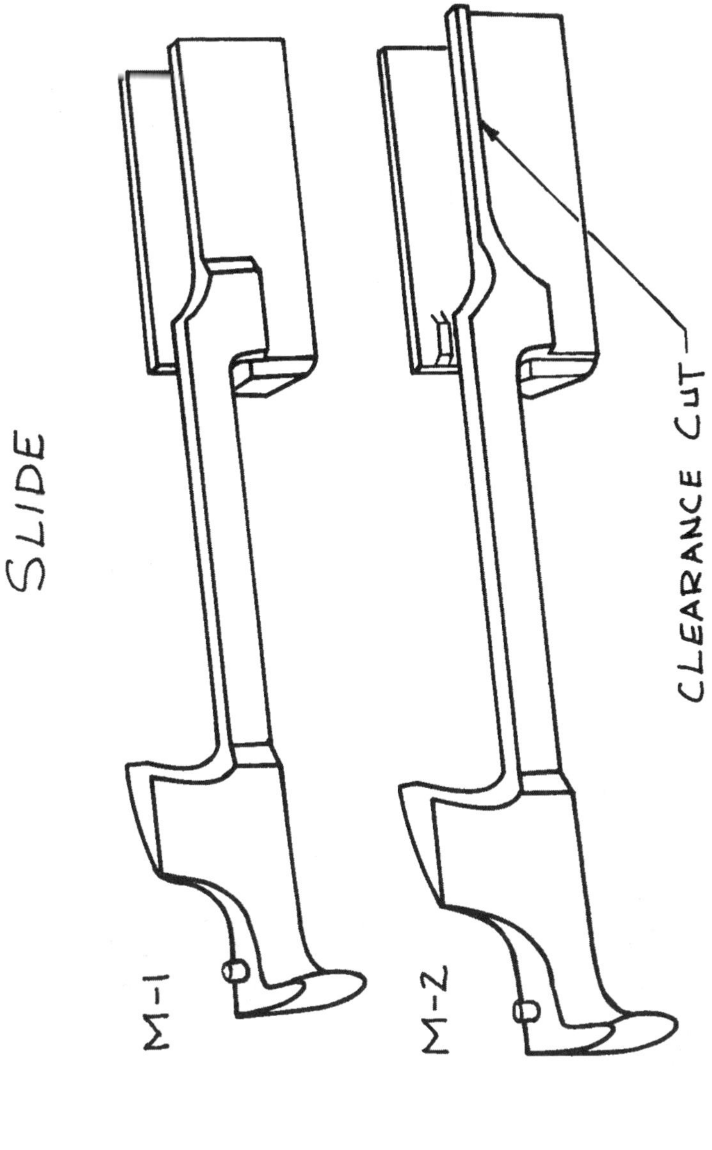
SLIDE
M-1
M-2
CLEARANCE CUT

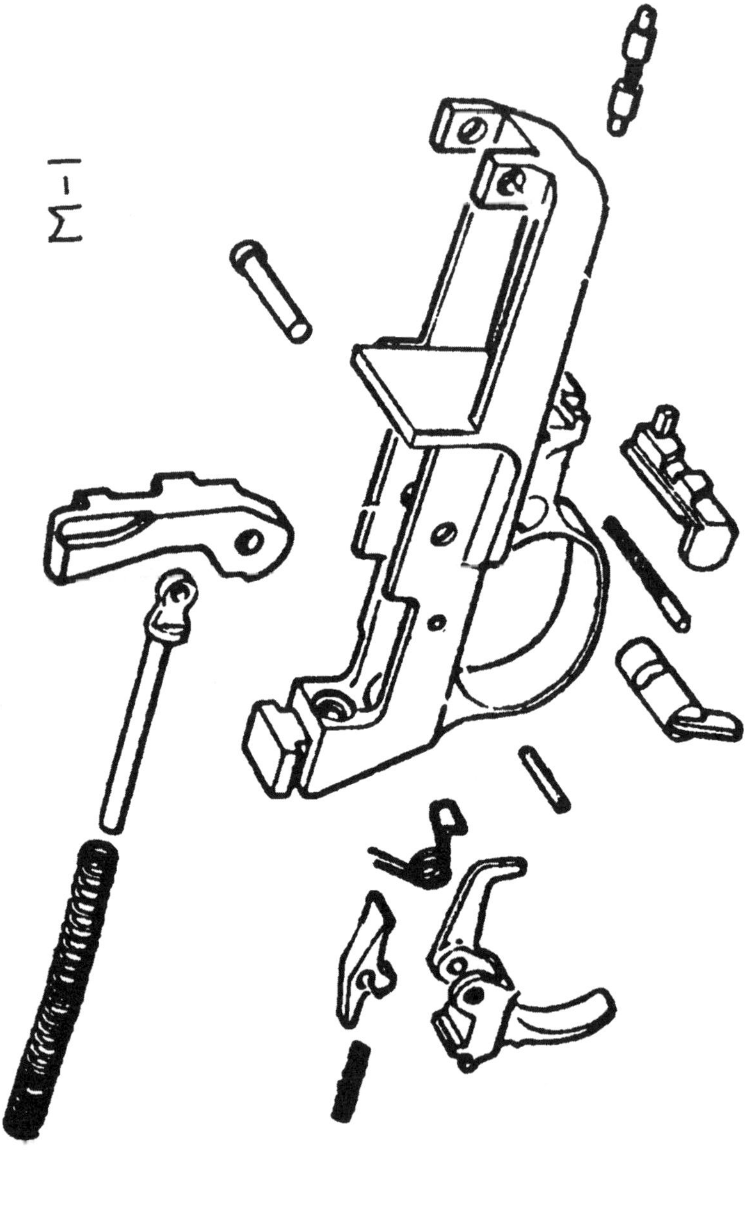
M-1

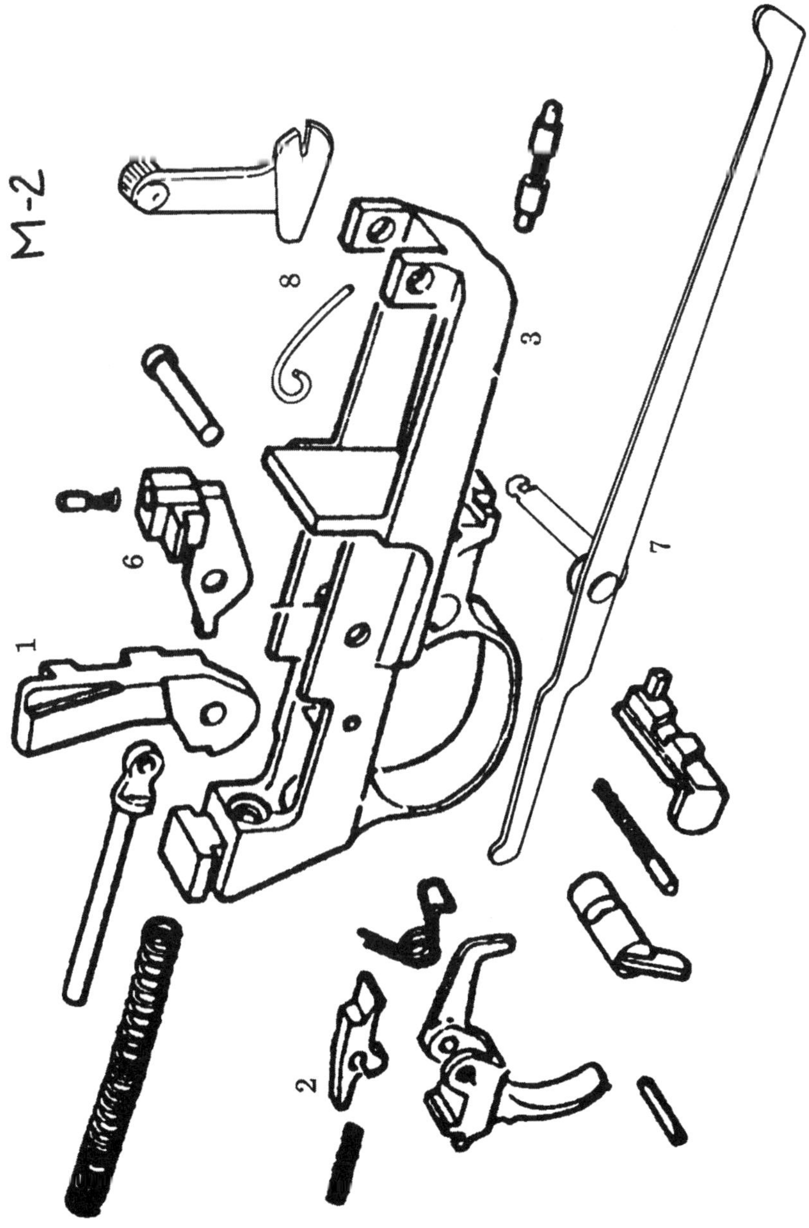
M-2
1
2
3
6
7
8

Notes

Notes

Other Books Available From Delta Press

01-20

081 Crossbows/ From 35 Years With the Weapon......$13.95
206 CIA Field Expedient Preparation of Black Powder......$21.75
221 Evaluation of Improvised Shaped Charges......$17.95
411 Homeopathic First Aid......$12.95
432 Cold Weather Survival......$16.30
435 Homestead Carpentry......$16.30
444 Leadership Hanbook of Small Unit Ops......$5.00
454 Survival Childbirth......$13.95
C-002 How To Open A Swiss Bank Account......$10.95
C-011 Defending Your Retreat......$10.95
C-020 Methods of Long Term Storage......$10.95
C-023 Federal Firearms Laws......$13.05
C-024 Select Fire Uzi Modification Manual......$15.21
C-025 How to Build Silencers......$10.95
C-027 Firearms Silencers Volume......$15.21
C-029 M1 Carbine Arsenal History......$8.67
C-032 The Silencer Cookbook......$10.95
C-038 How To Build A Beer Can Morter......$5.40
C-040 Criminal Use of False ID......$10.95
C-045 Improvised Weapons of American Underground......$16.30
C-046 Improvised Munitions Handbook......$14.12
C-047 Improvised Munitions Handbook......$14.12
C-049 The 007 Travel Kit......$13.03
C-050 Surviving Doomsday......$18.48
C-056 Navy Seal Manual......$32.65
C-061 MAC 10/11 Submachine Gun......$10.95
C-065 Poor Man's James Bond Vol 1......$38.10
C-066 The Anarchist Cookbook......$32.65
C-067 Terrorist Explosive Handbook......$21.75
C-114 AR-15, M16 and M16A1 5.56mm Rifles......$15.21
C-117 Ruger Carbine Cookbook......$15.21
C-120 Shootout II......$16.30
C-122 AK-47 Assault Rifle......$17.39
C-123 Combat Loads for Sniper Rifles......$20.66
C-125 Browning Hi-Power Pistols......$15.21
C-126 UZI Submachine Gun......$15.21
C-127 FullAuto Vol 1 Ar-15 Modification Manual......$13.03
C-128 Full Auto Vol 2 Uzi Mod Manual......$13.03
C-130 Full Auto Vol 4 Thompson SMG......$13.03
C-131 FullAuto Vol 5 M1 Carbine to M2......$13.03
C-135 M-14 Rifle, The......$17.39
C-136 Fighting Garand Manual......$17.39
C-137 Ranger Training & Operations......$21.75
C-138 Emergency War Surgery......$32.65
C-140 Emergency Medical Care/Disaster......$16.30
C-143 Survival Medicine......$17.39
C-150 Improvised Munitions Black Book Vol 1......$19.57
C-151 Improvised Munitions Black Book Vol 2......$19.57
C-152 Impro. Munitions Black Book Vol 3......$32.65
C-155 Fighting Back on the Job......$16.30
C-162 Colt .45 Auto Pistol Manual......$15.21
C-165 USMC Sniping......$19.62
C-168 Suppressors Vol. 1 For Ruger MK......$13.03
C-169 Suppressors Vol. 2 For Ruger 10/22......$13.03
C-107 Suppressors Vol. 3 For AK......$13.03
C-171 Suppressors Vol. 4 For Uzi......$13.03
C-172 Suppressors Vol. 4 For MAC 10/11......$13.03
C-175 Beat the Box......$8.67
C-197 How to Make Disposable Silencers Vol. 1......$21.75
C-198 Expedient Hand Grenades......$23.93
C-199 U. S. Army Counterterrorism Training Manual......$10.90
C-203 Explosives and Propellants......$21.75
C-204 Improvised Munitions/Ammonium Nitrate......$21.75
C-205 Two Component High Explosives Mixtures......$21.75
C-207 CIA Improvised Sabotage Devices......$21.75
C-208 Special Forces Demolition Handbook......$21.75
C-210 Guerilla Warfare......$17.95
C-211 FN-FAL Auto Rifles......$23.93
C-218 Professional Homemade Cherry Bomb......$17.39
C-219 Improvised Shaped Charges......$21.75
C-220 mprovised Explosives, Use In Detonation Devices......$21.75
C-227Springfield '03 Rifle Manual......$17.39
C-229 Lock Picking Simplified......$5.45
C-230 How to Fit Keys by Impressioning......$5.45
C-231 Lockout -Techniques of Forced Entr......$17.39
C-235 How to Open Handcuffs Without Keys......$17.39
C-244 Camouflage......$16.30
C-274 M14 and M14A1Rifles and Rifle Marksmanship......$26.11
C-284 Thompson Submachine Guns......$21.75
C-289 M1 Carbine Owners Manual......$17.39
C-290 Clandestine Ops Man/Central America......$16.30
C-324 How to Make Disposable Silencers Vol. 2......$21.75
C-337 Sten MKII SMG Construction Manual......$27.30
C-365 Survival Gunsmithing......$16.30
C-372 Poor Man's James Bond Vol 2......$38.10
C-386 Self-Defense Requires No Apology......$17.39
C-41 Select Fire 10/22......$17.39
C-415 Construction Secret Hiding Places......$17.39
C-416 Brown's Alcohol Motor Fuel Cookbook......$27.20
C-417 Canteen Cup Cookery......$10.95
C-418 Improvised Batteries/Detonation Devices......$21.75
C-421 Walther P-38 Pistol Manual......$15.21
C-422 P-08 Parabellum Luger Auto Pistol......$16.30
C-427 How to Train a Guard Dog......$18.48
C-429 The Anarchist Handbook Vol. 1......$16.30
C-430 Beretta - 9MM M9......$15.95
C-432 Mercenary Operations Manual......$16.30
C-439 Improvised Lock Picks......$19.57
C-455 Modern Day Ninjutsu......$19.57
C-477 Rifleman Went to War......$32.65
C-512 FullAuto Vol 8 M14A1 & Mini 14......$13.03
C-518 L.A.W. Rocket System......$5.45
C-539 Dental Specialist......$32.65
C-545 Advanced Homemade Fireworks......$21.75
C-558 Food. Fur and Survival......$21.75
C-561 The Squeaky Wheel......$5.45
C-562 Survival Chemist Book......$19.57
C-564 Medical Specialist......$32.65
C-571 U. S. Marine Corps Scout/Sniper Training Manual......$30.40
C-610 Aunt Bessie's Wood Stove Cookbook......$10.95
C-622 Defensive Shotgun......$16.30
C-625 MAC-10 Cookbook......$16.30
C-627 Brown's Lawsuit Cookbook......$29.38
C-628 Can You Survive......$18.48
C-633 Hand to Hand Combat......$13.03
C-634 USMC Hand to Hand Combat......$10.95
C-635 Hand to Hand Combat by D'Eliscu......$13.03
C-639 Police Karate......$19.57
C-640 Knife Throwing Techniques of the Ninja......$17.39
C-646 Science of Revolutionary Warfare......$18.48
C-668 Survival Shooting for Women......$16.30
C-680 Infantry Scouting, Patrol. & Sniping......$18.48
C-694 CIA Field Expedient Incendiary Manual......$21.75
C-761 How to Lose Your X-Wife Forever......$27.20
C-763 Vigilante Handbook......$21.95
C-775 Alcohol Distillers Handbook......$30.48
C-776 How to Build a Junkyard Still......$17.39
C-777 Engineer Explosives of WW......$21.75
C-794 US Marine Corps Essential Subjects......$21.75
C-815 US Marine Bayonet Training......$5.45
C-818 The Poisoner's Handboo......$32.65
C-829 The Anachist Handbook Vol. 2......$16.30
C-889 Napoleon's Maxims of War......$13.02
C-890 Company Officers HB of Ger. Army......$21.75
C-891 German Infantry Weapons Vol 1......$10.90
C-896 Op. Man. 7.62mm M24 Sniper Weapon......$5.45
C-899 US Army Bayonet Training......$5.45
C-9002 German MG-34 Machinegun Manual......$21.75
C-9012 Firearm Silencers Vol 3......$16.30
C-9013 HK Assault Rifle Systems......$30.48
C-9047 SKS Type of Carbines, The......$19.57
C-9048 Rough Riders, The......$27.20
C-9052 Private Weaponeer, The......$19.57
C-9055 Full Auto SKS Modification Book......$17.39
C-9057 Keys To Understanding Tubular Locks......$13.03
C-9060 The Anarchist Handbook Vol. 3......$16.30
C-9082 Lasers & Night Vision Devices......$32.65
C-9084 Assorted Nasties......$32.65
C-9101 Guide to Germ Warfare......$19.57
C-9102 CIA Field Expedient Method for Explosives Preparation......$21.75
C-9119 Surviving Global Slavery......$18.48
C-9120 Military Knife Fighting......$14.95
C-9127 H&R Reising Submachine Gun Manual......$21.75
C-9131 Live to Spend It......$32.65
C-9134 Military Ground Rappelling Technique......$17.39
C-9135 Smith & Wesson Autos......$21.75
C-9137 Caching Techniques of U.S. Army Special Forces......$16.30
C-9138 USMC Battle Skills Training Manual......$49.95
C-9139 Survival Armory......$30.48
C-9156 Concealed Carry Made Easy......$19.50
C-9164 The L'il M-1, The .30 Cal. M-1 Carbine......$19.57
C-9170 Urban Combat......$30.48

870-862-3811
info@deltapress.com

Desert Publications
505 East Fifth St., Dept. - BK - 166
El Dorado, AR 71730
www.deltapress.com

Shipping & Handling
1 item $7.95 - 2 or more
$12.95